CHRYSLER 300
1955-1961

John Gunnell

First published in 1992 by Motorbooks International Publishers & Wholesalers, PO Box 2, 729 Prospect Avenue, Osceola, WI 54020 USA

© John Gunnell, 1992

This is a reissue of the 1982 original edition, with corrections and revisions added

All rights reserved. With the exception of quoting brief passages for the purposes of review no part of this publication may be reproduced without prior written permission from the Publisher

Motorbooks International is a certified trademark, registered with the United States Patent Office

The information in this book is true and complete to the best of our knowledge. All recommendations are made without any guarantee on the part of the author or Publisher, who also disclaim any liability incurred in connection with the use of this data or specific details

We recognize that some words, model names and designations, for example, mentioned herein are the property of the trademark holder. We use them for identification purposes only. This is not an official publication

Motorbooks International books are also available at discounts in bulk quantity for industrial or sales-promotional use. For details write to Special Sales Manager at the Publisher's address

Library of Congress Cataloging-in-Publication Data
Gunnell, John.
    Classic motorbooks Chrysler 300 1955-1961 / John Gunnell.
      p. cm.—(Classic motorbooks photofacts)
    Rev. ed. of: Chrysler 300, 1955-1961. ©1982.
    Includes bibliographic references (p.   ) and index.
    ISBN 0-87938-643-6
    1. Chrysler automobile—History. I. Gunnell, John. Chrysler 300, 1955-1961. II. Title. III. Title: Chrysler 300 1955-1961 photofacts. IV. Series.
TL215.C55G87   1992
629.222'2—dc20   91-48467

On the front cover: The 1961 Chrysler 300 G owned by Bill Craffey of California. *David Gooley*

Printed and bound in the United States of America

ACKNOWLEDGMENTS

The author wishes to sincerely thank the following people, in alphabetical order, whose direct or indirect contributions made the creation of this book a reality: Ed Aldridge, Bev Aldridge, James Banach, Ray Beaumont, George Berg, Madeline Berg, James H. Brown, Michael A. Carbonella, Ron Chuchola, Henry Austin Clark, Jr, Linda Clark, George Cone, Carol Cunningham, Gil Cunningham, Diane Davis, Donna DeButts, Duane DeButts, Ray Doern, Jerry Elam, Merryanne Elam, Harry Ewert, Linda Ewert, Chester Foth, Gary Goers, Phil Hall, Bruce Hoover, Tony Hossain, Andy Jugle, Carl Kiekhaefer, Chic Kramer, Richard Langworth, Harold Longworth, Arnold Lueth, Bob McAtee, Evaline McAtee, Bill McBride, Terry McTaggert, Alan Moon, Gloria Moon, Eleanor Riehl, George Riehl, John Sheets, Don Warnarr, Jack Wiltse, Paul Yount, R. Perry Zavitz and Gregg Ziegler. To these and all other persons who have contributed to my love and understanding of the Chrysler 300 Letter Car I extend my heartfelt thanks.

It would certainly be an oversight to pass over two additional names, both of which deserve recognition in any literary effort devoted to the Chrysler 300. Bob Rodgers and Virgil Exner are the men I refer to—the men who conceived and brought to life the fabulous 'Beautiful Brute.'

A very special extra note of thanks should again be afforded to Harry Ewert, of Elmhurst, Illinois. Over the years of restoring his own 1959 300-E, Harry compiled a notebook containing a large quantity of information on the Letter Cars. In addition while President of the Chrysler 300 Club, Inc. he prepared, with the help of members of that organization, a series of Judging Criteria Sheets that provided most of the detailed information included herein. The utilization of this data will certainly be of major significance to all Chrysler Letter Car enthusiasts from this time on.

The major portion of the photos in this book were taken by the author. The cars photographed were selected on the basis of originality and/or quality of restoration and on the basis of their location within a suitable geographic area determined by limits of the author's personal travel. Photos contributed by other sources are captioned with suitable credit lines. In the few cases where nonoriginal features are seen on a car, this fact has been noted.

In any book of this sort there is a possibility of inadvertent mistakes. The author will welcome any documented evidence of such, but assumes no liability for the unavoidable occurrence of error. Restorers of Chrysler 300's would be best advised to use this book as a guide and to also become a member of one or both clubs devoted to the marque.

Prices shown for the various car models included in the text are based on a number of sources and reflect the recommended price of the model at a specific point in time. They are for the basic car without optional extras. The shipping weights quoted for any car may vary slightly from source to source depending upon manufacturer's weighing methods and the time and place the weight was recorded.

During the 1950's and 1960's Chrysler Corporation and Chrysler or Chrysler-Imperial Division made frequent concessions to customer demands for special equipment or finish. It is quite possible that some Chrysler 300's were built which did not conform to regular specifications or equipment offerings.

PREFACE

The Chrysler 300 Letter Car was offered as a limited production American automobile for eleven years between 1955 and 1966. A total of 16,857 examples were built. About ten percent of those cars remain today. All are considered collectors items. Those built through 1961 are recognized as "modern classics" and, thus, are certified as Milestone cars by The Milestone Car Society.

The first of the series was built in 1955 as an experimental model incorporating the most advanced thinking of Chrysler Corporation engineers. The car was immediately impressive and the company—in need of something to compete with *image* cars like the Corvette and Thunderbird—decided to make some available on a limited-production basis.

All 300's shared a number of common features which created the image of a powerful, full-sized, sports touring automobile. Their styling was a refined blend of assembly line elements custom-finished and appointed for enthusiast appeal. To this were added a race-bred, high-performance V-8, heavy-duty suspension, specially engineered drive train and competition-type running gear.

Besides being exclusive, Chrysler 300's were exceedingly fast and driveable. In 1955, Letter Cars were seen at Daytona Beach, where they took first and second in the Flying Mile and swept the field in the 160-mile Grand National stock car race. For the next ten years both of these events would be almost totally dominated by the legendary Letter Cars.

This book is an attempt to capture, on paper, the history and the heartbeat of the Chrysler 300. In a basic sense, it is merely a collection of facts about this great series of cars and will, hopefully, prove useful to collectors and restorers. But, it was prompted by a love of the Chrysler products in my past and I sincerely hope that this emotion will be reflected in the final product.
John Gunnell

CONTENTS

CHAPTER ONE
The Joy of Winning, Chrysler Racing History 6

CHAPTER TWO
1955 Chrysler C-300, The Car That Swept Daytona 12

CHAPTER THREE
1956 Chrysler 300-B, America's Most Powerful Car 20

CHAPTER FOUR
1957 Chrysler 300-C, America's Greatest Performing Car 28

CHAPTER FIVE
1958 Chrysler 300-D, Respectable Race Car 37

CHAPTER SIX
1959 Chrysler 300-E, The Lion-Hearted Letter Car 46

CHAPTER SEVEN
1960 Chrysler 300-F, Sixth of a Celebrated Breed 54

CHAPTER EIGHT
1961 Chrysler 300-G, Grand Touring Car, Letter Style 65

CLUBS 73
RESTORATION AIDS 74
INFORMATION AND SOURCES .. 76
INDEX 78

CHAPTER ONE
The Joy of Winning, Chrysler Racing History

When Lee Petty's gray Chrysler New Yorker came barreling through the north turn at the Daytona Beach race course in 1954, those watching knew that a major feat had been achieved by America's third-ranked auto maker. For the first time in many years Chrysler Corporation was making a car capable of winning important races.

Some enthusiasts were convinced Chrysler had forgotten entirely about its racing heritage, which was sad indeed. The company's formal beginning dates back to 1925, when it succeeded Maxwell Motor Corporation. Even Maxwell had built racing cars.

An early Maxwell had challenged the Pennsy Railroad's speedy Congressional Limited during 1905, in a run from Washington, D.C., to Baltimore. It lost by only four minutes. Others had competed at Ormond Beach and Sheepshead Bay Speedway. In 1906, a racing tradition was established when an experimental twelve-cylinder Maxwell appeared at Daytona Beach.

Chrysler's predecessor employed the services of some famous speed demons. Ray Harroun, Ralph DePalma and Eddie Rickenbacker were all company drivers. Harroun, winner of the very first Indy 500

The radical-looking Chrysler Airflow failed to gain public acceptance and cast a gloom over the company's sales in the thirties. Several of the streamlined cars set speed records at Daytona Beach.

The experimental Newport dual-cowl phaeton was built by Chrysler and appeared in 1940. Only six were made and one paced the 1941 Indy 500, becoming the first nonproduction car to have this honor. *(Roth-Handle Rarities Show)*

race, had even designed several kerosene-burning Maxwell race cars. One went 105 mph at the Indianapolis Motor Speedway in 1914. Another ran in the Elgin Road Race with Billy Carlson at the wheel.

Early *Chryslers* also held promise as race cars. In 1924 the company's famed "Three Musketeers"—engineers Fred M. Zeder, Carl Breer and Owen Skelton—created a 4.7:1 compression power plant that was considered a high-performance engine in its day. In 1924, at Mount Wilson, California, Ralph DePalma set records with a Chrysler. He returned, twelve months later, to do even better. In 1927, a quartet of Imperial roadsters competed at Le Mans, taking third- and fourth-class wins in the French Grand Prix. These cars, along with several Model 72 Chryslers, also finished strong in the 1928 Belgian Grand Prix.

In 1929, the roadsters were back at Le Mans; and a Chrysler-powered machine raced 89.63 mph, in America, to earn a qualifying berth at the Indy 500. Two years later, a stripped Model 77 roadster and a Locke-bodied Imperial appeared at the Swiss Grand Prix. Meanwhile, back home, another Model 77 set a New York-to-Palm Beach speed/distance record and the George N. Howie Special, with Howie driving, qualified for Indy at 102.84 mph. Howie finished eleventh. In the same year, Harry Hartz and Bill Arnold set records, at Daytona, with a heavy Imperial Eight sedan.

Such competition continued into the early 1930's, spurring increased sales and technical improvements, but the Great Depression struck Chrysler hard. Then came the failure of the Airflow program, provoking an immediate cutback in creative engineering and competitive efforts. One of the radical-looking streamliners became the last factory-supported race car to appear for some twenty years. This was a coupe with which Harry Hartz established seventy-two Daytona stock car records, including a blistering 95.7 mph Flying Mile run.

After the failure of the Airflow, Chrysler racing was dead. A few private drivers, like Fred Agabashian, kept the nameplate alive in local events until 1948, when a new type of racing grew popular. Sanctioned by groups like the American Automobile Association (AAA) and the National Association for Stock Car Automobile Racing (NASCAR), late model stock car rac-

The development of the Chrysler Fire Power hemi engine paved the company's way back into competition. It was introduced as the standard V-8 for the 1951 Saratoga and New Yorker models.

ing was initially dominated by Oldsmobile Rocket 88's and six-cylinder Hudson Hornets.

The year 1948 also brought the development of an exciting new experimental Chrysler power plant. This was the hemi V-8, which evolved from sixteen years of research and concentrated wartime efforts. As early as 1935, Chrysler had toyed with the L-head engine alternatives. Initial work was on straight sixes and eights then in regular use. Around the time that World War II broke out, company engineers built a double-overhead-camshaft six-cylinder engine with a spherical segment cylinder head. It seemed most promising. Hemispherically shaped combustion chambers had been seen in aircraft engines as early as 1918 and were used by Stutz, Duesenberg, Miller and Offenhauser for racing. The principle was well-suited to air-cooled aero engines, but was first considered unacceptable for automotive use.

When hostilities ended, the project picked up again. The overhead camshafts were dropped and a new layout was emphasized. This V-8 configuration had better medium-speed volumetric efficiency and a shorter size. The second advantage made it more suitable to postwar styling trends. In late 1948, a prototype power plant was quietly constructed. Three years later, the new Fire Power V-8 was installed in regular production cars, like the stock car racer that Lee Petty took to Daytona.

The best way to illustrate the advantage of the hemi V-8 is to compare it with the comtemporary Cadillac engine. The two had exactly the same cubic-inch capacity, but different cylinder head designs. The Chrysler used the spherical segment heads and was rated at 180 hp. The Cadillac, with conventional wedge-shaped heads, was twelve percent less potent. Chrysler's spherical combustion chambers, with the spark plug in the center of the bore, generated less heat loss and featured wider-spaced, smoother and straighter valve ports. Professional racers quickly realized that the Fire Power V-8 was an excellent competition choice.

One of the first enthusiasts to use the hemi in competition was Briggs Cunningham, who had raced Le Mans in 1950 with two Cadillac-powered cars. For 1951, Cunningham switched to the new Fire Power motor with Chrysler providing some special hard-

The powerful New Yorkers earned the Ram-Charger nickname. Despite a large 131.5-inch wheelbase and weight of 4,500 pounds, the hemi-powered production car was a high-performance machine in this era. *(Old Cars Price Guide)*

ware. The result was a 300-hp V-8 quite like that used in later Letter Cars. Chrysler supplied the engine blocks to Cunningham at a forty-percent discount backed up with free, computerized technological advice. The blocks were installed in three roadsters dubbed Cunningham C-2's. At Le Mans, one of these cars finished eighteenth, while the others were sidelined by mechanical problems. Considering that they were pitted against Europe's finest factory works teams, it was quite an admirable performance.

The same year, another hemi racing effort was made by Carl Kiekhaefer, a boat motor manufacturer from Fond du Lac, Wisconsin. Kiekhaefer's Mercury Outboard Company, formed in 1939, had competed in hydroplane racing and set several water speed records. When the Power Boat Association issued a decree against racing, Kiekhaefer turned to automobiles. During 1950 he heard about the first running of the Carrera Panamericana, or Mexican Road Race, and decided to compete. Two Fire Power Chryslers were sponsored as Mercury Outboard entries for 1951.

The cars shared a number of modifications including two-barrel Zenith carburetors, Mallory ignition systems and special Mercury Outboard spark plugs. Common suspension alterations included stiff competition springs, dual telescopic shock absorbers and heavy-duty sway bars front and rear. Special oversized fuel tanks were used, along with a novel braking system that relied on dry ice for cooling. The first car, driven by John Fitch, had an over-bored hemi with solid valve lifters. Mechanical ills retired it from the race. Tony Bettenhausen piloted the second Chrysler—with stock bore and stroke and hydraulic tappets—into sixteenth position.

Adding more spark to Chrysler's track performances during 1951 was the appearance—at the Daytona Speed Trials—of a Chrysler with automotive writer Tom McCahill at the wheel. He was road testing the car for *Mechanix Illustrated* and decided to take it racing. With little more than a careful tune-up, "Uncle Tom" hit the sand with his showroom stock New Yorker and racked up a two-way run at an average of 100.13 mph. McCahill was named stock car

The Chrysler C-200 was a Ghia-built show car with convertible styling. Touches like the wide wheel cutouts, front and rear design treatments and chromeplated wire spoke wheels were added to the production Letter Cars.

class champ and Chrysler reaped suitable publicity.

As a total entity, Chrysler's 1951 stock car racing remained just a token effort. A first-place win in a 250-mile Grand National race at Detroit was recorded. Yet, it was quite clear that the hemi, with only minimal work, had the potential to be a real winner. Not until the 1955 Chrysler 300 appeared, would this potential be fully realized.

By 1952, more hemi engines were running races in America and overseas. The Cunningham team returned to Le Mans with a 600-pound-lighter C-4 model. One coupe and a pair of roadsters, these cars had 300 hp and five-speed gearboxes. Only one finished, but most impressively: It earned fourth place with an average speed of 87 mph.

In Mexico, one of Kiekhaefer's 1951 New Yorkers, with Roger McFee at the wheel, finished fifth amid controversy. Kiekhaefer argued that the Lincoln team, sponsored by Ford Motor Company, had made some illegal modifications. Meanwhile, on the Daytona Beach oval, Pat Kirkwood's Chrysler ran fastest at 110.97 mph, but a Hudson proved a better long-distance machine in the main event. The Fire Power V-8 seemed like an engine with potential, but the competition was getting bigger and better. Cadillac and Lincoln had moved to four-barrel carburetion, while Hudson and Oldsmobile were offering racing parts as factory accessories.

For the 1953 French Grand Prix, Briggs Cunningham came up with a 310-hp C-5 roadster model, one of which finished third with Fitch and Phil Walters driving. A C-4 roadster took seventh and a C-4 coupe finished tenth. Late in the year a specially equipped New Yorker DeLuxe was released by the factory. Intended primarily for racing, a total of fifteen such cars were built over two years. The 1953 editions had dual four-barrel carburetors and a 280-degree roller tappet camshaft. Their 235-hp output rating was an inspiration to Chrysler enthusiasts.

A more significant development of 1953 was a spin-off of Cunningham's efforts. Four exotic hemi engines were designed for the Indy 500. They had massive internal modifications and produced 400 hp on alcohol fuel. Based on a preliminary AAA ruling legalizing stock-based motors up to 5.5 liters (335 cubic inches) for the Memorial Day classic, one of the hemis was

The people who collect Chrysler Letter Cars today are rarely adverse to fast driving. In 1979 the members of the Chrysler 300 Club International took a spin around the high-speed oval at Chrysler's Chelsea Proving Grounds.

bolted into a Kurtis chassis for prerace testing. It ran 135 mph for 140 miles straight. In further trials the car went a distance of 900 miles without a spark plug change. Owners of Meyer-Drake-engined cars panicked and pressured the sanctioning body to revoke its displacement rule change. The Chrysler engines were quickly destroked to 271 cubic inches and revised with equipment including ram stack Hilborn fuel injectors, roller tappets, 300-degree cams, large valves, forged rocker arms, magneto ignition and a 13.0:1 compression ratio.

In this form the engines were capable of 372 brake horsepower (bhp) at 5800 rpm, but lost a certain amount of torque advantage needed for Indy. The cars did not race, however, and the motors and two chassis were sold to Firestone Tire and Rubber Company for testing purposes. The Ohio-based firm still owns the engines today.

For the 1953 Mexican Road Races, Chrysler decided to go a bit further than ever before, offering a heavy-duty engine and suspension package. It included a higher compression ratio, roller tappets, specially ground cam, dual-quad intake manifold, Air-Lift suspension stiffeners, front and rear stabilizer bars, heavy-duty axles and Imperial disc brakes. Even more significant was the fact that Chrysler's chief engineer, Bob Rodger, traveled south of the border to observe the race.

This brought a corporate reassessment of strengths and weaknesses in the Chrysler racing program. For 1954, the Indy fiasco was written off as a bad experience and factory involvement was deemphasized for Le Mans. The stock-based 235-hp specials were continued, but now had single four-barrel carburetors and electric fuel pumps.

In late summer, to christen Chrysler's Chelsea Proving Grounds, Tony Bettenhausen led a team which drove a New Yorker over the high-speed oval for an entire day. It averaged 118 mph for twenty-four hours straight. A number of high-output sedans were given a special 300-plus-hp *export* package for Mexico, but the race was canceled.

Since standard transmissions would not be introduced for Chrysler Letter Cars until the 1956 300-B models came out, all these racers used an automatic transmission—really beefed up! In contrast to street models, the racing examples had higher compression, roller-tappet cams, anti-roll bar, "air-lift" suspension and heavy-duty rear axles. For better cooling and strength most included Kelsey-Hayes Imperial wire wheels; also Imperial disc brakes. Unofficially a race version had over 300 hp and could hit 135-140 mph!

In February 1955 Brewster Shaw and Lee Petty took checkered flags. By the fall of the year an unusual car was seen moving down the line in Chrysler's Jefferson plant. It was not an Imperial, like other cars in the same row. It was the mockup for a new type of machine . . . the 1955 Chrysler C-300. This was the first of the fabulous Letter Cars!

CHAPTER TWO
1955 Chrysler C-300
The Car That Swept Daytona

The first of the fabulous Letter Cars to appear in Chrysler show rooms was called the C-300. It represented a brutally different kind of American car in the mid-fifties. The C-300 was built as a limited-production model with European looks, race car power, sports car agility and country club appointments.

This great machine evolved from four factors influencing Chrysler's destiny in late 1954: the market failure of the current line, the growing realization that Chrysler products were stylistically out-of-date, a new awareness of the hemi V-8's exciting performance potential and growing competitive pressure to create a special model built for sport and speed.

Chief research engineer Bob Rodger gets credit for the Letter Car concept and is called the "father of the 300." He moved up to the top engineering post in 1952, after work on hemi development. Despite his clean-cut corporate image, Rodger loved the flamboyant world of racing and his Mexican visit, during 1953, started a dream.

In August 1954, Rodger presented his thoughts about a semi-stock, sports/performance coupe to Chrysler President Ed Quinn, who was ready to listen.

The optional wire wheels were the same used earlier on Imperials and the manufacturer was Kelsey-Hayes. Exner liked them. Exhaust pipe deflectors were standard, but often deteriorated. This car shows a nonoriginal type. *(Tony Hossain)*

Quinn knew that something had to be fielded against the likes of the Chevrolet Corvette and Ford Thunderbird. Serious discussions brought design chief Virgil Exner into the picture. Exner said such a project might be practical if the special features for the car could be adapted without new tooling. Quinn agreed and Rodger got the okay to formulate rough plans and cost estimates.

A hand-picked group of men—all speed enthusiasts—was assembled to work on the project. Rodger was the driving force, but Exner grew more enthusiastic. The designer sketched a proposal with sculptured sides and a wide, horizontal-bar grille. Meanwhile, stylist Cliff Voss joined the team and Tom Poirier came aboard from production. A few other stylists and engineers were enlisted, almost by personal invitation. By late September, some basic decisions had been made. Most stemmed from the monetary limitations governing the job. The sculptured look and bar grille were out, so a sketch of a New Yorker body with an Imperial grille was substituted. Exner favored a car with almost no surplus chrome trim.

Engineering was based on off-the-shelf hardware including three main items: a racing-type 300-hp hemi engine, the Mexican Road Race "export suspension" package and a beefed-up automatic transmission. By October, a 1955 New Yorker two-door hardtop was wheeled into the Jefferson plant and stripped of engine, trim and interior fittings. It then became the prototype for the first Letter Car, as it was totally reassembled on an Imperial line in a hand-built manner. Parts were taped on to finalize trim arrangement. The result was a New Yorker with Imperial looks up front and a Windsor appearance at the rear. The racing engine under the hood could barely maintain low idle, but at top end it would go like a rocket.

As this car rolled down the line it drew cheers from an excited crowd of employees, and company executives soon got the chance to drive it. According to Chrysler's *Spectator*, "It rode like a truck and idled rough, but it had the maneuverability of a polo pony and the speed of wind."

By November, word was given to start production and the Automobile Manufacturers Association (AMA) was contacted to register the new model. Dur-

Fitting an Imperial grille to a Chrysler body took some work, but made the C-300 quite distinctive. (Prototypes had an Imperial bumper up front, but this treatment did not make production.) The front of the car had simple lines with minimal ornamentation including the Chrysler name and a checkered-flag emblem between the grilles.

The Chrysler C-300 was the company's first mid-year model. Each example got an extensive three-hour road test prior to being approved for delivery. Ingredients included trim features from several models and a race-bred hemi V-8. This 1955, owned by Bob McAtee, has only 28,000 miles on it.

A pop-up air intake was incorporated at the cowl of the C-300. It had a wire-mesh screen to keep debris from the air ducts. The windshield had bright stainless steel reveals. Electric wipers and a windshield washer system were standard on the C-300. Solex tinted glass and power window lifts were options shown here.

ing December, a few finishing touches were added. Chrysler parking lights were substituted for Imperial units up front. A special 150-mph speedometer was added and the Imperial front bumper was replaced by a Chrysler type to lighten the appearance. Changes were made almost until the time production began.

On January 17, 1955, the C-300 was announced to the public in a statement carefully timed to precede the Speed and Performance Trials at Daytona Beach. Chrysler microfilm records show that the lowest serial number car (No. 3N551001) was built on February 10, 1955, and shipped six days later. Legend has it that the first car was painted Platinum White and the next ten Tango Red, but Chrysler microfilm proves this mostly incorrect. Seven of the first ten *were* Platinum White, two were Black and only one was Tango Red. It's also legend—and probably true—that some very early cars were built with irregular trim attachments. The only upholstery combination provided was tan leather.

The appearance of the C-300 was a credit to Exner's abilities. Walt Woron, of *Motor Trend*, spoke of its Italian flavor. While all 1955 Chryslers had some European influence, the Letter Car was a purebred when it came to looks. There was an obvious and strong resemblance to experimental show cars which Chrysler had built, by Ghia in Italy, during the early fifties. It was hardly surprising that the show car names, such as K-310, were closely copied for the first Letter Car, as it was truly a special automobile. The looks were not vastly different from other 1955 Chrysler coupes, but many little changes added up. Touches like a clean-shaven hood with no ornament and wide-spaced Imperial bumper guards, up front, made the model distinctive. Large expanses of unadorned sheet metal set off every badge and band of chrome like a sparkling jewel.

The frontal treatment had a simple elegance combined with a purposeful look. There was nothing extra and nothing missing and contemporary enthusiasts couldn't help but appreciate such simplicity on a mid-century MoPar. In profile, the C-300 seemed shorter than it was. It had a plain look that was rounded, but crisp at the same time. Front fenders were tall and the entire body swept to a rearward taper. Viewed from the side, the C-300 had a wedge-

The wide chrome headlight bezels were peaked at the top. Very early 1956 cars used chrome bezels, too, but later cars had the same part finished in body color.

Most cars used an Imperial wheel disc modified with a checkered insert having a gold 300 insignia. Wire wheels were an option and came as a set of five. Remington L78x15 tires are a good approximation of the originals.

The C-300 had semi recessed-type door handles adapted from Ghia-built Chrysler show cars. A round lock with a swiveling cover was positioned directly below the handle. Opening the doors was accomplished by the method seen here. The side windows had weather-resistant stainless steel garnish moldings. Serial numbers were located at the front door hinge post.

shaped look, appearing wider in back than in front. From most rear angles the roof looked much lower than it actually was, a pleasant optical illusion. Both the hood and deck lid seemed to blend into the overall design like interlocking puzzle pieces. The C-300 lacked extraneous ornamentation and had only a minimum of badges and emblems. There was no provision for outside rearview mirrors and these items were not on the factory-approved accessory list. In some cases, they *were* added at the dealer level. Only a handful of factory options and accessories was available, however. Included were power steering, power seat, electric clock, radio and antenna, heater, Solex tinted glass and electric window lifts. Imperial wire wheels were offered, at extra cost, because Exner liked them and there were several hundred leftover sets in factory storerooms. Normally, Imperial wheel covers were used, but had a checkerboard center with gold 300 identification added.

Some one-off variations from equipment specifications listed in factory literature were made. For instance, it has been reported by several sources that a casino in Nevada asked for, and received, a Chrysler 300 (Imperial) station wagon. It had the New Yorker body, an Imperial front end and the C-300 trim and special hemi engine.

Above all else, the C-300 was a race-bred car. Originally there were only vague notions about incorporating distinctive styling and never was appearance a primary concern. The Letter Car program was much more focused on engineering and with bringing Chrysler Division's best go-fast technology into the show rooms. It exploited the untapped reservoir of power in the hemi V-8, which had become apparent through racing. Chrysler, in fact, was very lucky that the car came out looking so good, since there had never been a clay model. The mockup assembled at the Imperial factory was built and trimmed by trial and error.

The basis for the Letter Car power plant was the standard 331-cubic-inch-displacement hemi block. It received a full-race camshaft with long-duration valves that increased the peak power range by 600 rpm. The cam was ground to provide above-average performance. Twin Carter WCFB carburetors were mounted on special intake manifolds designed to en-

The 300's had crisp-looking bumpers with a massive guard protecting the license plate. The exhaust deflectors seen here are the original type and they are very hard to find today. This car was featured in a parade prior to the 1979 World 600 stock car race at the Charlotte Motor Speedway.

The left rear fender on the C-300 was decorated similar to the right side, but did not have a gas filler door. The overall level of body decoration was kept to a minimum. Rear quarter moldings were adapted from Windsors. While the chromed taillamp housings did not blend into the fender line their bright finish made them seem unobtrusive. Chrysler enthusiasts do not call them fins. Wheel cutouts were on the full side for 1955.

courage optimum breathing. Solid lifters rounded out the package, which had the same 8.5:1 compression ratio as New Yorkers. Dual exhausts were bolted to the engine and had pipes measuring one-quarter-inch more in diameter than the New Yorker type. With all of this equipment a rating of 300 bhp resulted. It was the first American *production* car to churn out so much power and the model designation emphasized this fact. (Note: The Duesenberg SJ was actually America's first 300-hp car, but cannot be considered a "production" auto.)

The only transmission used on 1955 Letter Cars was the two-speed PowerFlite, with modifications including a higher stall-speed torque converter to accommodate the extra torque of the engine. There were five rear axle gear ratios available. Specially constructed high-speed tires added to performance. They had a lower rate of horsepower absorption than normal types allowing up to twenty-five extra usable horses at 100 mph. At this point, a Letter Car would still be *accelerating*. The C-300 topped out at around 130! A well-tuned example could move from zero-to-sixty in about ten seconds and cover the standing-start quarter-mile in under eighteen seconds, reaching approximately 82 mph. This makes it seem like there's no top end; even though Tim Howley, writing in *Special-Interest Autos,* described the shifting as jerky, steering as heavy, the idle rough and the ride firm.

As expected, the C-300's turned up at Daytona in 1955, competing in Class Four, for cars of unlimited displacement. The effort had a good deal of factory backing and received the support of supplier firms, too. For example, Auto-Lite spark plug engineer Paul Atwell did much work prepping the cars. Also seen at Daytona was a team of Chryslers campaigned by Carl Kiekhaefer, who was still determined to have the public associate the name of his firm with the best engineering around. He decided to sponsor a fleet of Chrysler 300's in both big-time stock car racing associations: NASCAR and AAA. His Mercury Outboard Motor Team was assembled for the sole purpose of winning checkered flags.

The best drivers of the day were hired to pilot the Kiekhaefer cars. Tim Flock, Buck Baker, Speedy Thompson and other well-known aces would all be

The emblem seen at the rear deck matched that on the front of the car. It showed a gold 300 insignia on a black-and-white-checkered field. The round key lock for the trunk was located directly below the badge.

The seats (optional four-way electric shown here) were upholstered in tan natural cowhide leather with pleats on the inner cushion panel running widthwise, fully across. The lower panels on the split seat backs repeated this look. Pleats on the seat cushion front bolster and top of the backrest ran lengthwise. Black carpeting was used on all cars.

employed at one time or another. There were several fleets of cars, about twenty-five top mechanics and some $100,000 worth of equipment involved in the effort. New drivers were often recruited right on the sands. Flock, for example, was visiting Daytona after a self-imposed exile. (He had won the championship in 1952, but had hotly contested a 1954 disqualification that sent him packing for home.) A Kiekhaefer representative hired Flock on the spot, just prior to the 1955 races. It was a good move.

By the time Speed Week was over, the C-300's had broken just about every record on the books. They took the NASCAR Flying Mile championship with Warren Koechling's 127.58-mph run. The Standing Start Mile also went to a Letter Car with a speed of 76.84 mph. But, Flock took home the biggest win of all with his performance in the Daytona Grand National. He set a qualifying speed record of 130.293 mph and took the checkered flag, in the 160-mile main event, with an average speed of 92.05.

Daytona was just the beginning. Tim Flock would win thirteen consecutive races behind the wheel of a C-300 in the following months, netting him the NASCAR Grand National Stock Car Championship. In AAA racing, Frank Mundy piloted another Kiekhaefer Letter Car to numerous victories and the title. Overall, thirty-seven major races of the season went to the C-300's. It was the first time in history that one make of car captured top honors in both leagues.

SPECIFICATIONS
Model number   C-300
Serial number data   (location) front door hinge post (start) 3N55-1001 (end) 3N55-2725
Engine number data   (location) top front of block behind water pump (start 3N55-1001 (end) 3N55-2756
Start production date   February 10, 1955
Announcement date   January 17, 1955
Advertised dealer price   (hardtop coupe) $4,055.25

The rear seat upholstery had a pattern similar to that used in front. The tan headliner was bordered with bright metal reveals around window openings and a grooved trim plate on the roof pillar. A circular courtesy lamp with white plastic lens and wide chrome trim ring was set into the roof pillar. A plated coat hook was seen above the rear quarter window.

Door panels had a tan, smooth-grained covering and a wide strip of black carpet along the sill. Thin stainless moldings arranged in "tree branch" pattern were on the upper door panel. A lever-action handle opened the door from inside.

Though not listed under options and accessories in the C-300 sales catalog, a rear radio speaker was available as a dealer-installed item. Proper mounting was on the tan vinyl-covered package shelf, as shown here on McAtee's car.

Standard equipment

Special hemi engine, heavy-duty suspension, PowerFlite automatic transmission, windshield washer, custom steering wheel, dual-exhaust system, padded dashboard, power brakes, custom interior and exterior appointments and trim.

Options and Accessories
(Code) — description — price
- (8) — power steering — $113
- (4) — four-way power seat — $70
- (2) — electric windows — $102
- (8) — standard-tune radio and antenna — $110
- (1) — Touch-Tone radio and antenna — $128
- (4) — heater — $92
- (1) — Solex tinted glass — price not available
- (8) — Kelsey-Hayes wire wheels — $617.60 (set of five)
- (*) — 3.36:1, 3.73:1, 3.91:1 and 4.1:1 axles — no additional charge
- (1) — electric clock — price not available

Original paint colors

(factory code)	name	Ditzler combination
(01)	Black	9000
(25)	Tango Red	70525
(30)	Platinum	8096

Interior trim
Standard upholstery color — Tan
Standard upholstery material — Leather

Engine
Type — 90-degree Fire Power V-8, overhead valves, cast-iron block
Bore x Stroke — 3.81x3.63 inches
Mains — five
Displacement — 331 cubic inches
Brake hp — 300@5200 rpm
Torque — 345@3200 rpm
Compression — 8.5:1
Camshaft drive — chain type

Valves
Intake head diameter — 1.94 inches
Exhaust head diameter — 1.75 inches
Intake stem diameter — .372 inch
Exhaust stem diameter — .372 inch
Overall length (both) — 5-1/32 inches
Cam duration — 280 degrees (intake) 270 degrees (exhaust) 60 degrees (overlap)

Carburetion
Type — (2) Carter WCFB four-barrel
Carburetor models — (front) WCFB-2318S (rear) WCFB-2317S
Carter repair kit — 1833A (original number)
Carter gasket kit — 282A (original number)
Exhaust system type — dual exhausts
Muffler Type — reverse-flow

Electrical
Original battery type — 6-volt Auto-Lite 2H-135-RD (optional) Willard MW-2-135-R
Original starter — Auto-Lite No. MCL-6121-A
Original generator — Auto-Lite No. GGW-6001
Original generator (power steering) GGW-6016
Original voltage regulator — Auto-Lite VBE-6001-A

Transmission
Type — PowerFlite two-speed automatic; torque converter with gears
Ratios — (low) 1.72:1, (drive) 1.00:1, (reverse) 2.39:1
Lever location — instrument panel to right of steering wheel

Running gear
Differential — hypoid drive type
Axle gear ratios — (standard) 3.54:1 (optional) see accessory list
Standard manual steering — worm-and-roller type with 5.5 turns lock-to-lock and overall ratio of 20.4:1. Turning circle is 46 feet 6 inches wall-to-wall
Optional power steering — Chrysler full-time coaxial type with 3.5 turns lock-to-lock and overall ratio of 16.2:1
Standard power brakes — Chrysler adjacent type, hydraulic internal-expanding

Dash lever for two-speed PowerFlite transmission protruded from slotted plate to right of engine-monitoring gauge cluster dial. The radio, when ordered as on McAtee's car, was positioned to the right and heater controls were mounted in separate pods.

Evaline McAtee demonstrates method of shifting gears on the C-300 via the dash-mounted control lever. The rearview mirror was hung from the top center of the windshield header. No outside mirrors were provided.

A black dash pad was standard. The right-hand side of the dash was painted Saddle Beige and housed an electric clock, glovebox and radio speaker grille. A Chrysler script was seen as decoration.

with vacuum assist. Front and rear linings (primary and secondary) are 12.57x2x.20 inches. Swept area is 201 square inches

Chassis and body
Frame type   welded, double-channel box section with side rails and lateral cross-members
Body construction   welded steel
Wheelbase   126 inches
Front tread   60.2 inches
Rear tread   59.6 inches
Overall length   218.6 inches
Overall height   60.1 inches
Overall width   79.1 inches
Ground clearance   5.8 inches
Shipping weight   4,005 pounds
Fuel tank capacity   20 gallons

Suspension
Front, type   independent, lateral control with coil springs
Spring rate   800 pounds
Rear, type   parallel-set, 7 semi-elliptic leaves
Spring rate   160 pounds

Cooling system
Radiator type   fin-and-tube down-flow
Coolant capacity   25 quarts (without/heater) 26 quarts (with/heater)

Wheels and tires
Tire size   8.00x15
Type   4-ply Goodyear Super-Cushion Nylon special
Whitewall width   approximately three inches
Standard wheels   15x5.5K 5-lug steel disc type
Optional wheels   Kelsey-Hayes chrome wire wheels

Production totals
Two-door hardtop coupe   1,725
Two-door convertible   not offered
Total   1,725
Remaining 1981*   136 (7.8 percent)
Notes: No C-300's were built with standard transmission. All cars had tan leather interior. Several C-300 "replica" convertibles have appeared in recent years, but are not factory-built units.

*All figures for the number of cars remaining in 1981 is from the Chrysler 300 Club International's serial number *Register*.

Shop manual - The 1955 Chrysler shop manuals were printed before the Letter Car was introduced. They are usable, but do not apply specifically to this model. Bob Rodger provided personal mimeographed service bulletins and notes to owners on an individual request basis.

Sales literature
1—Folder No. CS-339 2/55, 4 pages, color, 11 x 9 inches, *The Chrysler 300*.
2—Sheet, unnumbered, 1 page, black & white, 8½x11 inches. Lists race records and specifications.
3—Facts book, No. (C), 9 pages, 5 x 7 inches. Published by Ross Roy, Inc. Lists complete specifications and comparisons with other 1955 American cars. Looseleaf form with various pages added at different dates during 1955.

Advertisements
1—*Motor Trend*, June 1955, page 3.
2—*Mechanix Illustrated*, May 1955, page 191
3—*Mexhanix Illustrated*, August 1955, page 23
4—*Popular Mechanics*, September 1955, page 239
5—*Motor Life*, May 1955, page 7
6—*Sports Car Illustrated*, August 1955, page 23

The 1955 Fire Power V-8 used a unique triangular air cleaner and large valve covers that were finished in gold. The black bar, attached by screws to the center of the valve cover, provided access to spark plugs.

Two Carter four-barrel carburetors were standard equipment on the C-300 power plant. At Daytona, Kiekhaefer took dirt samples from engine oil after each race or run and analyzed them. This led to development of special air cleaners.

CHAPTER THREE
1956 Chrysler 300-B
America's Most Powerful Car

The second Chrysler 300 appeared about twelve months after the C-300. It was called the 300-B and was again based on the New Yorker body with an Imperial grille, heavy-duty suspension and high-performance hemi V-8. Styling and power refinements were seen, resulting in modernized looks, added muscle under the hood and a more civilized demeanor. The majority of styling changes were shared with other Chryslers and technical alterations centered on increased engine displacement, push-button automatic transmissions and the introduction of floating shoe, center-plane brakes. A new 340-hp base engine made the 300-B the most powerful car available in America.

The option list also grew this year. The C-300 had been provided with only a minimum of extra equipment offerings, although priced rather high. Consequently, many buyers who thought they were purchasing a powerful *luxury car* had been disappointed. The Letter Car was conceived and constructed essentially as a high-performance machine and it lacked many features and characteristics that buyers in this price range expected. Thus, power windows, air conditioning and self-winding clocks mounted in the

The second Letter Car looked much like the original edition and was again based on the Chrysler New Yorker body shell fitted with an Imperial grille, special trim, heavy-duty suspension and a race-bred hemi V-8. This Regimental Red example is owned by Jim Brown of Pennsylvania.

center of steering wheels were made available for 1956, in an attempt to improve sales. Because of such refinements, the 300-B is, today, generally regarded as a slightly more desirable car than the original C-300. Collectors prefer the second-edition Letter Car because of a cleaner appearance, engineering improvements, the additional options most cars were sold with and their increased rarity, due to lower production.

When the regular 1956 Chrysler line was introduced on October 21, 1955, the design of the 300-B was still being finalized. It was still considered a limited-production model and scheduled for mid-year introduction to focus attention on it. The C-300 had been successful in creating a great deal of prestige for the company and a January introduction was one way to keep the excitement level high. Only one car was built to preliminary 1956 specifications before late December, when ten more examples were made. The new car seemed to reflect a greater emphasis on aircraft styling with its modest, fin-shaped rear fenders. Under the hood was, of course, the most powerful motor in the industry.

On January 4, 1956, the official announcement of the new model was made. A premier showing took place two days later at the annual Chicago Auto Show. The 300-B was on display in the windy city through January 15 and appeared in dealer show rooms shortly thereafter. Many books say that the car was the same as the C-300, which is fine until you start hunting for parts. A look at a Chrysler manual will tell you that the mufflers, speedometer heads, tail pipes and many other parts for each year were not the same.

Styling, however, was basically unchanged in many regards except toward the rear of the car. The back end treatment was greatly enhanced with larger, finned fenders producing a dart-like overall shape. Taillamps were now vertically incorporated into the fin towers with a large red lens sitting on top of a smaller white plastic unit that comprised a backup lamp. The fins and bumper angled toward the front of the car, with the latter continuing as a large bar stretching across the back of the car from end-to-end. The frontal design was identical to that of the 1955 Letter Car in most details. Grilles were the same

The real news about the 300-B was found under the hood. A larger, more powerful base engine made it the best-performing car in America. In addition, optional high-horsepower heads were released late in the year. Cars with the extra-cost V-8 were the first American production units to exceed 1 hp per cubic inch, an honor which Chevrolet later claimed undeservedly.

chromed Imperial type with a retouched emblem positioned on the painted nose panel between them. There was now a stainless steel strip that ran from side-to-side, below the grille and behind the bumper, and the headlamp bezels were done in body color.

As seen from the side view, obvious differences for 1956 included the shape of the finned rear fenders, the slanting rear body line and the relocation of some minor trim features. A few cars were built with a snubbed-off belt molding that fell short of the taillight housing. Another varying feature was the finish around the taillight lenses, which was stainless steel on some cars and painted silver on others. As in 1955, only three paint combinations were available and all interiors were upholstered in tan leather with black carpeting on the floor and the bottom third of the tan door panels.

From a technical standpoint, the 300-B was designed to be even more competitive in racing than the C-300. A more powerful 355-hp engine option and TorqueFlite automatic transmission were introduced. In addition, a number of cars were built with stick-shift attachment. They could go from 0-60 in just 8.2 seconds and had a top speed approaching 135 mph.

The standard 300-B engine was the 354-cubic-inch Fire Power V-8. It was good for 340 hp, or more with a new option—three-inch-diameter header pipes—added. The compression ratio was raised by milling the heads slightly over 1955, but most other goodies were the same as the year before. The new motor did, however, get a specially hardened crank, malleable-iron main bearing caps and tri-metal bearings.

Featuring the same bore and stroke, the optional high-horsepower motor came with a 10.0:1 compression ratio. This was reported to be the result of using thinner head gaskets, but actually the 355-hp heads were milled even more than the standard hemi heads for 1956. The three-inch exhaust system was standard on this motor and peak horsepower came at 5200 rpm. Some sources indicate it was provided prior to the 1956 Daytona Speed Week, which cannot be documented. AMA specifications were updated to re-

High-angle front view shows that the headlight rims are now painted body color and trimmed with a thin molding on the front edge. Some very early production units may have had the chome 1955-style bezels.

Front end styling was similar to 1955 with small detail changes. The inside of the chromed grille was finished in silver paint. The upper section of the model badge had a band of black behind the 300 number designation. The lower section had a black-and-white checkerboard field against which was placed the letter B, trimmed in red.

flect this option on April 16, 1956, but it was not until June 13 that Chrysler released a confidential bulletin announcing the high-horsepower heads.

After offering just one transmission for 300's in 1955, two more were added during 1956. Early in the year the PowerFlite two-speed automatic was still standard equipment, but gear selection was switched to push-button control. A high-performance torque converter and planetary gear set were incorporated. Also available throughout the year was a three-speed, column-mounted manual gearbox pirated from the Windsor line and beefed-up with a special eleven-inch-diameter clutch having twenty-three percent more effective area. A total of thirty-one stick-shift cars were built and had a number of common alterations. A black 1953 Chrysler steering wheel was used on these cars and the left and center portions of the dashboard were taken from 1955 Chryslers that did not have push-button shift controls. The slot for the 1955 shifter was covered with a chrome plate. The standard transmission and linkage were identical to the parts used on 1956 Dodges, and power brakes were not offered for cars with this setup.

Late in 1956, the cast-iron, three-speed TorqueFlite automatic transmission was added to the 300 spec sheets. This was a unit of much better design and greatly enhanced the performance of the cars, especially between 0-30 mph. The extra gear was a big help in getting the B underway a bit faster. This was the only year that Letter Cars were offered with three different gearboxes.

Common comments about the 300-B included the opinion that it idled even rougher than the 1955 edition, but shifted much smoother. An interesting note is that with the car (with automatic) in neutral, the handbrake must be on—there was no "park" gear. Stock '56's seemed to ride somewhat softer than '55's. And the car seemed to start off slower than you would expect; but still *accelerating* at 80-90 mph—it's a good thing the brakes this year were excellent.

Tim Flock, Carl Kiekhaefer, the Mercury Outboard racing team and dozens of Chrysler 300-B's showed up at the Daytona Safety and Performance Trials. On February 28, 1956, Flock took one of the Mercury Outboard cars to a win in the 160-mile Grand Nation-

Full wheel covers of the Imperial type were standard equipment. They had a dished design, domed center and finned pattern on the outside portion. A round plastic checkerboard medallion was inserted at the center and carried the 300 number designation in gold.

The side view of the 300-B was dominated by the new rear end treatment with slanting rear body lines. There were heavy-duty Oriflow shock absorbers and high-rate seven-leaf springs to keep that heavy-looking bumper from sagging the rear of the car.

al stock car race. His average speed was 90.836 mph on a track heavily dampened by rain. In addition, he was able to set a new two-way record for the Flying Mile with a 139.37-mph run. It was an impressive mark that would stand until 1958, for all manufacturers, and two years more for Chrysler.

Shortly after Daytona, Carl Kiekhaefer pulled the Mercury Outboard fleet out of AAA sanctioned racing but continued to run in NASCAR competition. His 300's took first place in twenty-two out of fifty-six races during the year, with Buck Baker winning the driver's championship for the season. Notable Chrysler performances were recorded at Charlotte, Portland, Syracuse, Eureka (Kansas) and Martinsville (Virginia). By the season's end the boat motor magnate was fighting with NASCAR officials, too. Kiekhaefer was not about to stand for favoritism being shown to factory teams, especially that of Chevrolet. It was a successful year for 300 stock car racing, but it would also be the last time that Letter Cars competed in circle track racing on a substantial level.

SPECIFICATIONS
Model number 300-B
Serial number data (location) front door hinge post (start) 3N56-1001 (end) 3N56-2150
Engine number data (location) top front of block behind water pump (start) 3NE56-1001 (end) 3NE56-2174
Start production date December 1955 (one car built October 1955)
Announcement date January 4, 1956
Advertised dealer price (hardtop coupe) $4,312.25 (convertible) not offered
Standard equipment
Special hemi engine, heavy-duty suspension, PowerFlite automatic transmission, power brakes, custom steering wheel, Safety-Cushion dash panel, prismatic rear-view mirror, nylon racing-type tires with white sidewalls, chrome rear license plate frame, electric clock, windshield washer, undercoating, stainless steel wheel covers and hand-brake warning signal.
Options and accessories
(Code) — description — price
(*) — 3-inch-diameter exhaust system
(*) — 5.38:1, 5.83:1 and 6.17:1 rear axle ratios.
(*) — 15x9.5-inch steel disc wheels
(1) — manual shift transmission — $70

The spare tire sat upright in a deep well and the evaporator for the air-conditioning system was positioned behind the rear seat backrest. The trunk had gray carpets and neat sidewalls of heavy cardboard.

The scoops at the rear of the roof on this car are for air-conditioning ducts. The fins had a curving upper contour which blended into the body panels at the back edge of the door, imparting a kicked-up rear fender look.

(5) — air conditioning — $567
(1) — Electro Touch-Tone radio and antenna — price not available
(3) — Music Master radio with rear seat speaker — price not available
(1) — Solex tinted glass — price not available
(2) — Electric window lifts — price not available
(4) — 4-way power seat — price not available
(6) — Instant Heat heater — price not available
(4) — Custom Conditionaire heater — price not available
(3) — Self-winding steering wheel center clock — price not available
(9) — power steering (automatic transmission only) — $97
(*) — 355-hp Fire Power V-8
(*) — paper element air cleaner
(*) — 16.2:1 steering gear ratio
(8) — Kelsey-Hayes wire wheels — price not available
(5 & 7) — Hi-Way Hi-Fi (portable record player) — price not available
(*) — These are racing and performance options, prices and codes not available.

Original paint colors
(factory code)	name	Ditzler combination
(01)	Black	9000
(37)	Regimental Red	70643
(41)	Cloud White	8036

Interior trim
Standard upholstery color — Tan
Standard upholstery material — Leather

Engine
Type — 90-degree Fire Power V-8, overhead valves, cast-iron block
Bore x Stroke — 3.94x3.63 inches
Mains — five
Displacement — 331 cubic inches
Standard bhp — 340@5200 rpm
Optional bhp — 355@5200 rpm
Standard torque — 385@3400 rpm
Optional Torque — 405@3400 rpm
Standard compression — 9.0:1
Optional compression — 10.0:1
Camshaft drive — chain type

Valves
Intake head diameter — 1.94 inches
Exhaust head diameter — 1.75 inches
Intake stem diameter — .372 inch
Exhaust stem diameter — .372 inch
Overall length (both) — 5 1/32 inches
Cam duration — 280 degrees (intake) 270 degrees (exhaust) 60 degrees (overlap)

Carburetion
Type — (2) Carter WCFB four-barrel
Carburetor models (front and rear, early production) WCFB-2314SA, (front) WCFB-2444S (rear) WCFB-2445S (after 2/4/56)
Carter repair kit — 1832 (original number)
Carter gasket kit — 282 (original number)
Exhaust system type — dual exhausts (3-inch diameter optional at extra cost)
Muffler type — reverse-flow

Electrical
Original battery type — 12-volt Auto-Lite 12H-70 (S.A.E. type 3SM-70)
Original starter — Auto-Lite No. MDF-6001
Original generator — Auto-Lite No. GJC-7002
Original voltage regulator — Auto-Lite VBE-6001-A

Transmission
(POWERFLITE)
Type — two-speed automatic; torque converter with gears
Ratios (low) 1.72:1 (drive) 1.00:1 (reverse) 2.39:1
Lever location — push buttons, left-hand side of instrument panel
(TORQUEFLITE)
Type — three-speed automatic; torque converter with gears
Ratios (low) 2.45:1 (drive) 1.45:1 and 1.0:1
Lever location — push buttons, left-hand side of instrument panel

The rear quarter molding was longer this year and the lettering was positioned below the rear of the chrome strip. The red plastic portion of the taillights was shorter, but there were now white back-up lenses at the bottom.

Bumper guards for 1956 were of a cleaner design without a cross bar. They incorporated license illumination lamps. The same badge used between the radiator grills also appeared on the deck lid. Below it was the round face of the keyhole cylinder.

(STANDARD TRANSMISSION)
Type   three-speed manual
Ratios   (low) 2.50:1 (second) 1.68:1 (third) 1.0:1
Lever location   column-mounted lever with conventional pattern

Running gear
Differential   hypoid drive type
Axle gear ratios   (standard) 3.54:1
from New Yorker series   (optional at no charge) 3.36:1, 3.73:1, 3.91:1
from Imperial series   (optional at no charge) 4.10:1, 4.30:1, 4.56:1, 4.89:1
from Dodge truck   (optional at extra cost) see accessory list
Standard manual steering   (see 1955 specifications)
Optional power steering   (see 1955 specifications)
Standard power brakes   Chrysler floating-shoe, center-plane type, hydraulic internal-expanding with vacuum assist. Front and rear linings (primary and secondary) are 12.57x2.5x.20 inches

Chassis and body
Frame type   welded, double-channel box section with side rails and lateral cross-members
Body construction   welded steel
Wheelbase   126 inches
Front tread   60.4 inches
Rear tread   59.6 inches
Overall length   222.7 inches
Overall height   59.4 inches
Overall width   79.1 inches
Ground clearance   5.8 inches
Shipping weight   4,145 pounds
Fuel tank capacity   20 gallons

Suspension
Front, type   independent, lateral control with coil springs
Spring rate   800 pounds
Rear, type   parallel-set, 7 semi-elliptic leaves
Spring rate   160 pounds

Cooling system
Radiator type   full-length water-jacket cooling with thermostatic bypass temperature control and fin-and-tube downflow radiator
Coolant capacity   25 quarts (without heater) 26 quarts (with heater)
Fan type   4-blade (standard) 6-blade (with air conditioning)

Wheels and tires
Tire size   8.00x15, 4-ply
Type   4-ply Goodyear Blue Streak racing type
Whitewall width   approximately 3 inches
Standard disc wheels   15x6K painted black
Optional disc wheels   15x9.5L painted black
Optional wire wheels   Kelsey-Hayes wire spoke, chromeplated

Production totals
Two-door hardtop coupe   1,102
Two-door convertible   not offered
Total   1,102
Remaining 1981*   116 (10.5 percent)
Notes: Total number of stick-shift cars recorded at factory was 31 units.
*see Chapter Two

Sales literature
1—Folder No. CS-369, 1/56, 4 pages, color, 14x11 inches, CHRYSLER 300-B America's Most Powerful Car.
2—Facts Book, No. (C), 8 pages, 5x7 inches. Published by Ross Roy, Inc. Lists complete specifications for full line; some equipment comparisons with other cars; trim options. Looseleaf form with pages dated 9/55 and 1/56.

Advertisements
1—Cover drawing from sales folder. Performance-oriented. Various publications.

Interior appointments were similar to 1955. The sides and backs of the front seat were done in tan vinyl, as was the headliner and upholstery piping. The black, padded dash cushion covering had white stitching at the corner and Saddle Beige paint was used at the right side of the dash. The side and door panels were covered with tan vinyl with carpeted lower portions. The rear package shelf was trimmed in tan grained vinyl. There were chrome inside windshield moldings and gray rubber pedal pads.

Automatic transmission was standard and had push-button controls at the extreme left of dash. Three different gearboxes were used in 1956 production. Cars with stick-shift attachments got certain specific interior alterations. Optional here are four-way power seat, power steering and power windows.

Jim Brown's 300-B was seen at the multi-club meet at the Chrysler Proving Grounds during the fall of 1979. A drive over the test track's high-speed oval was part of the event. His car has an optional heater and push-button radio.

In 1956 the accessory list grew and more gadgets went under the hood. This warranted 12-volt electrics to carry the additional load. Finish for some electrical components included silver generator brackets and black-colored spark plug covers, generator, starter and voltage regulator. The spark plug wire loom was wrapped in black plastic tape. This car carries the optional TorqueFlite transmission and air conditioning.

Cornering is no problem in an early Letter Car thanks to special springs, shocks and sway bars adapted from Mexican Road Race cars. The ride is firm, but most corners can be taken at even keel.

CHAPTER FOUR
1957 Chrysler 300-C
America's Greatest Performing Car

New sheet metal, more powerful engines, torsion-bar front suspension and the expansion of paint colors and body-type availability characterized the Chrysler 300 series for 1957. The letter C was back again, but was now moved behind the numerical designation. The new Letter Car was called the 300-C.

Previous Letter Cars had been customized Chryslers with few truly unique parts, but for 1957 Virgil Exner was charged with designing the first of the series ever to be built from scratch. He had already given free reign to his imagination and created a styling concept, called the "613," which was made into a full-size mockup. This was his idea of a male-oriented sports car and, to varying degrees, influenced the look of all 1957 Chrysler products. The 300-C was most deeply affected.

The "613" program began in October 1955, and by that November was brought to a highly finished stage with paint and foil trim. The front of the car was a dead ringer for what became the 1957 Letter Car. Luckily, some rear end features were changed when the styling of two show cars—designed by Chrysler's Maury Baldwin—provided a secondary design influ-

The 300-C had a handsome profile with its tapering, Flight Sweep-inspired fins and scooped-out wheel openings. Another Letter Car legend has it that the new-model medallions were patriotically colored to emphasize the American roots of the world-renowned Chrysler performance car. *(Chrysler)*

ence. The Flight Sweep I was a beautiful, Baldwin-inspired convertible and the Flight Sweep II his sporty coupe.

On December, 8, 1956, the new Chrysler 300 made its debut at the New York Auto Show. An overall cleanness of line, a more dramatic dart shape, an air-scoop-style snout and Ferrari-inspired grille made it clear that this was a car for the enthusiast buyer. It was to become the best-selling of all Milestone Letter Cars.

Appearance highlights of the 1957 model included a low, sleek body that looked somewhat larger than it was. The driving position seemed lower and more natural; the seats were at a slightly different angle, and the steering wheel seemed smaller. This was still a New Yorker-based variant on a 126-inch wheelbase and it was only one-half-inch longer than the original Chrysler 300. The front carried a large, trapezoid-shaped egg-crate grille as its most prominent trademark. This was surrounded by a "steer horn" bumper and functional brake-cooling ducts. The profile featured horizontal character lines, with tall, broad rear fins. Wheel openings were large, wide and low and, once again, a bare minimum of trim was used to decorate the body. Another new beauty treatment that became a Letter Car standard was a large, red-white-and-blue model identification medallion that was speared, through its center, by a horizontal rear quarter molding.

An entirely new roofline, for the hardtop coupe, had thinner arch-shaped pillars combined with a larger backlight and a vast wraparound windshield. Total vision area for this model was 4,601 square inches. And there was now a Chrysler 300 convertible.

At the rear was a long, flat and wide deck with a corner-bent back panel housing an indented license plate recess. Tall fin towers surrounded vertical red plastic lenses set on top of smaller backup lamps. The rear bumper was a straight, three-section affair without guards.

Simplicity and purity of line keynoted the overall design. If anything, there was less chrome than before and a totally integrated look. Five exterior finishes were provided and tan leather interiors were standard. A handful of cars were ordered with interior combinations not officially listed as options. These

The Chrysler styling staff was intrigued by aircraft motifs as reflected by these two 1955 show cars: the Flight Sweep I and Flight Sweep II. Features such as tailfins and general cleanness of line had a secondary influence on the appearance of the production 300-C.

units received an "888" code designation which signified a *nonstandard* upholstery selection, rather than a specific color choice. Such cars are exceptional rarities today.

In order to maintain its status as America's most powerful production car the 300-C got an enlarged hemi engine. Despite a heavier weight and slightly longer 219.2-inch-overall length, the Letter Car coupe was capable of moving from 0-60 in just 7.8 seconds with the least powerful engine. A high-horsepower option was also available. With this extra-cost V-8 fitted, the 300-C matched all cars but the Corvette in performance, and featured superior handling as well. Steering, however, was slower than it should have been unless power assist was added. The car had a hard ride, but was still far better than most performance cars of the day including the Studebaker Hawk, Plymouth Fury and Corvette. The 300-C was the only full-size, six-passenger American car capable of *cruising* at 100 mph. Its air-duct brake-cooling system and new torsion-bar suspension were among the most important technical innovations in the industry this year.

Torsionaire-Ride was hailed as a revolutionary system with torsion-bar front springs and ball-joint wheel suspension. Handling was much improved with these additions. Briggs Cunningham had used a similar setup on his 1953 Le Mans machinery and all Chryslers got it for 1957. The 300's system was special with forty-percent-stiffer bars to provide the race car type of ride. In combination with several other factors, like revised front-end geometry and a low center of gravity, the Torsionaire feature produced the best road handling in America at the time.

The standard 300 engine for 1957 was a 375-hp job with special hardware including mechanical tappets, stronger push rods, adjustable rocker arms, double valve springs, valve seat inserts, twin four-barrel intake manifold and carburetion, dual air cleaners, lighter-weight valves, hardened crankshaft, tri-metal bearings, high-output camshaft, special piston rings, custom calibrated distributor, high-temperature spark plugs and speed-limiting radiator fan.

An optional 390-hp engine-and-chassis package provided a more radical camshaft, 10.0:1 compression; limited-slip differential; 2.5-inch, low back-

New for 1957 was the 300-C with frontal styling derived directly from the "613" experimental prototype. The oblong, egg-crate grille was often referred to as "brutish-looking," which later moved automotive writer Bill Carroll to coin the description "Beautiful Brute." This convertible owned by George Berg is equipped with all accessories except air conditioning, and powered by the base Fire Power V-8.

Outside rearview mirrors were options this year. Flashing around headlights was usually painted body color, but may have been finished in silver for some cars. The lights and brake scoops were set into housings banded with thin metal moldings, which really stand out on this Jet Black car. Five solid colors were listed in the Letter Car sales catalog this season.

A small, red-white-and-blue plastic medallion was housed at the top center of the grille surround and had a chrome 300 insignia. Round badge was selected, according to Chrysler, since shape was unique to the automobile. Collectors use "medallion" exclusively to refer to these emblems. New bumper, chrome dust (gravel) shield and parking lamps tucked into grille were other frontal changes.

pressure exhaust system; heavy-duty clutch and driveshaft, *manual* transmission and *manual* steering. This package was primarily for the professional racers. Also offered was a choice of thirteen different axle ratios that allowed performance to be tailored to any conditions of use.

Two transmission choices were available. The three-speed TorqueFlite automatic was *standard* equipment. A three-speed manual gearbox was part of the chassis-and-engine option package. Power steering, power brakes and air conditioning were not available on manual transmission attachments. The automatic had push-button control and the governor on the unit was recalibrated for the higher torque output of the 392-cubic-inch motor. On the eighteen cars built with stick-shift, a 1957 Windsor steering wheel was used and a cover, made of ashtray insert material, was placed over the push-button control housing.

Due to a competition ban instituted by the AMA and a number of NASCAR rule changes, Chrysler 300 racing tapered off for 1957. The Daytona Safety and Performance Trials were the sole exception. Driver Red Byron captured first place in the Flying Mile at 134.128 mph. The 300-C was not quite as fast as the lighter B, but it was still damn fast! Using a fifth wheel and electric speedometer, Ross Roy, Inc.—the Chrysler advertising agency—timed a 375-hp coupe at the Chelsea Proving Grounds. It clocked 0-60 in around 8.6 seconds with a number of 0-100-mph runs in the 24.4-second range. The testing showed the car's top speed was 145.7 mph. Another magazine claimed a 0-60 time of 7.7 seconds and a quarter-mile acceleration run of seventeen seconds at 84 mph. In 1957, the 300-C was undoubtedly "America's Greatest Performing Machine," although it was not the fastest Letter Car and most of its performing was done on the streets.

SPECIFICATIONS
Model number   300-C (Chrysler C-76)
Serial number data   (location) front door hinge post   (start) 3NE57-1001   (end) 3NE57-3251
Engine number data   (location) top front of block behind water pump (start 3N57-1008 (end) 3N57-3338

There was a tapering "fin" running up center of the hood. Power-operated top was standard on convertible and came in black or ivory colors with either stitched or laminated seams. The bows below the canvas (Hartz cloth) were finished in Sahara Tan.

The brake-cooling air scoops had red-painted grilles. The fiberglass ducts and extensions were finished to match. The cool-air system worked well and gave the 300-C good deceleration characteristics in an era when most U.S. cars could go, but not stop well.

Start production   December 1956 (3 cars built October 1956)
Announcement date   December 8, 1956
Advertised dealer price   (hardtop coupe) $4,864 (convertible) $5,294

Standard equipment
Special hemi engine, heavy-duty suspension, custom steering wheel, Safety-Cushion dash panel, prismatic rearview mirror, nylon racing-type white sidewall tires, electric clock, directional signals, hand-brake warning signal, power brakes (automatic transmission only), TorqueFlite transmission, chrome stainless steel wheel covers, windshield washer, undercoating, leather upholstery, dual headlamps (single headlamps in states where dual headlamps were not legal), Silent-Flite fan drive, rear license plate frame, dual air cleaners, foam-rubber seat cushions, twin four-barrel carburetion, carpeting, drip rail cover molding, remote-control interior light, luggage compartment light, map light, full-flow oil filter, spare tire cover and inside glare-proof mirror.

Options and accessories
(Code) — description — price
(344) — air conditioning
(319) — Music Master radio and antenna
(321) — Electro Touch-Tune radio and antenna
(322) — rear shelf radio speaker (hardtop only)
(351) — power radio antenna
(335) — left-hand outside mirror
(541) — right-hand outside mirror
(324) — rear window deflector (hardtop only)
(323) — Custom Conditionaire heater
(352) — Instant Heat heater
(*) — power steering (automatic transmission only)
(328) — power windows
(326) — 6-way power seat
(332) — Solex tinted glass
(348) — shaded backlight (hardtop only)
(353) — rear quarter panel stone shields
(*) — 5.38:1, 5.83:1 and 6.17:1 axles
(*) — engine-chassis performance package — $550.00
Note: Prices for all 1957 options would be close to or the same as the prices for similar features in 1958. Exact 1957 prices are not available. Technical options do not have normal accessory codes.

Original paint colors
(factory code)   name   Ditzler combination
(X)   Cloud White   8036
(P)   Gauguin Red   70693
(A)   Jet Black   9000
(F)   Parade Green   41826
(N)   Copper Brown   21018
Note: Code 999 indicates non-standard paint, any color, on special orders (Chrysler Corporation colors only). It is possible that non-Chrysler paint was also special-ordered. Non-Chrysler colors were uncodeable.

Interior trim
Standard upholstery color   Tan
Standard upholstery material   Leather with vinyl on front seatback
Seat welting, color/type   Tan vinyl
Dashboard   Sahara Tan
Manual seat track shield   Sahara Tan vinyl plastic
Power seat track shield   Black vinyl plastic
Package shelf   Black vinyl plastic
Carpets   Black loop pile

Convertible top
Colors   Black or Ivory
Material (per supplier)   Double-stitch-seam canvas or laminated-seam canvas

Engine
Type   90-degree Fire Power V-8, overhead valves, cast-iron block
Bore x Stroke   4.00x3.90 inches
Mains   five
Displacement   392 cubic inches
Standard bhp   375@5200 rpm

A steel "button," shaped like the five-pointed Chrysler star emblem, was mounted at the center of the hood crease and used as the nozzle for the windshield-washer system. The cowl vent was perforated bright metal. Radio antenna mounting was made at the trailing edge of the right front fender. On cars with right-hand outside mirrors, the antenna was ahead of the mirror.

The gas filler door was on the left rear fender. The chrome molding encircling the model medallion had short extensions at the three and nine o'clock positions. The horizontal quarter panel moldings then locked into the extensions.

Optional bhp   390 @ 5400 rpm
Standard torque   420 @ 4000 rpm
Optional torque   430 @ 4200 rpm
Standard compression   9.25:1
Optional compression   10.0:1
Camshaft   chain type
Valves
Intake head diameter   2.0 inches
Exhaust head diameter   1.75 inches
Intake stem diameter   .372 inch
Exhaust stem diameter   .372 inch
Overall length   5.09 inches (intake) 5.05 inches (exhaust)
Cam duration (standard V-8)   280 degrees (intake) 270 degrees (exhaust) 60 degrees (overlap)
Cam duration (optional V-8)   300 degrees (intake) 300 degrees (exhaust) 95 degrees (overlap)
Carburetion
Type   (2) Carter WCFB four-barrel
Carburetor models   (front) WCFB-2534S (rear) WCFB-2535S
Optional carburetor type   (2) Holley four-barrel
Standard exhaust system (hardtop)   dual exhausts with 2¼-inch exhaust and extension pipes and 2-inch-diameter tail pipe
Standard exhaust system (convertible)   dual exhausts with 2¼-inch extension pipes and 2-inch exhaust and tail pipes
Optional exhaust system (hardtop)   dual exhausts with 2½-inch low back-pressure pipes throughout
Standard mufflers (hardtop)   one 35-inch muffler each side
Standard mufflers (convertible)   one 15-inch muffler each side with one 12-inch resonator each side
Air cleaner   dual paper-element type
Electrical
Original battery type   12-volt, Auto-Lite 12H-70
Original starter   Auto-Lite MDL-6001
Original generator   Auto-Lite GHM-6010-A
Air-conditioning generator   Auto-Lite GHM-6011-A
Original voltage regulator   Auto-Lite VRX-6201-A
Transmission
(TORQUEFLITE)
Type   three-speed automatic; torque converter w/gears
Ratios   (low) 2.45:1 (drive) 1.45:1 and 1.0:1
Control location   push buttons, left-hand side of instrument panel
(STANDARD TRANSMISSION)
Type   three-speed manual
Ratios   (low) 2.31:1 (second) 1.55:1 (third) 1.0:1
Lever location   column-mounted lever w/conventional pattern
Running gear
Differential   hypoid drive semi-floating type
Axle gear ratios   (standard) 3.36:1 (optional at no charge) 2.92:1, 3.18:1, 3.54:1, 3.73:1, 3.91:1, 4.10:1, 4.30:1, 4.56:1 and 4.89:1 (optional at extra cost) see accessory list
Standard manual steering   worm-and-roller type with 5.2 turns lock-to-lock and an overall ratio of 29.9:1. Turning circle measures 49 ft. 5 in. wall-to-wall and 43 ft. 11 in. curb-to-curb
Optional power steering   Chrysler Full-Time integral coaxial type with 3.3 turns lock-to-lock and an overall ratio of 19.8:1.
Power brakes (automatic only)   Chrysler Total-Contact type. Primary and secondary linings (front and rear) measure 12.6 x 2.5 x 13/64 inches. Swept area is 251 square inches. Duct cooling.
Chassis and body
Frame type   welded, double-channel box section
Body construction   welded steel
Wheelbase   126 inches

At the rear there were tall, red plastic taillight lenses and short, white backup lamps, which were both set into a common housing with chromeplated finish. Both red and white lenses were coded CHRC. Berg's car shown here won third and second place in 1977 and 1978 respectively at the Milestone Car Society Grand National.

Black-painted steel disc wheels were the sole offering this year and came in just one size. Polished chrome stainless steel wheel covers were standard. A dome with three screwed-on vanes, spaced 120 degrees apart, was either bolted or crimped to the covers. The center, where the dome was attached, was painted red.

The rear license plate was housed in a chrome-trimmed "shadow box" centered on the rear face of the deck lid. A three-piece back bumper was used. While possibly available as a dealer-installed accessory, bumper guards were not listed in the 300 folder. The rear bumper incorporated an upper bar that wrapped around the side of the body, a lower bar and a chromeplated dust shield. This last item was placed between the body and the upper bar. The fins angled slightly outward at the top. When Berg bought this car new in 1957 it had only 37 miles on the odometer—over 289,000 today!

Front tread   61.2 inches
Rear tread   60.0 inches
Overall length   219.2 inches
Overall height   54.7 inches (hardtop)
    55.0 inches (convertible)
Overall width   78.8 inches
Ground clearance   5.1 inches
Shipping weight   4,235 pounds (hardtop)
    4,390 pounds (convertible)
Fuel tank capacity   23 gallons
Suspension
Front, type   torsion-bar with ball joints
Spring rate   165 pounds
Rear, type   7 semi-elliptic leaves
Spring rate   135 pounds
Cooling system
Radiator type   full-length water-jacket cooling with thermostatic bypass temperature control
Coolant capacity   25 quarts with heater
Wheels and tires
Tire size   9.00x14
Type   Goodyear Blue Streak, whitewall, tubeless
Whitewall width   2.75 inches
Standard disc wheels   14x6.5
Production totals
Two-door hardtop coupe   1,767
Two-door convertible   484
Total   2,251
Remaining 1981*   291 (12.9 percent)

Notes: Total number of stick-shift cars recorded at factory was 18 units.
*See Chapter Two.
Sales literature
1—Folder No. CS-386, 12/56, 4 pages, color, 13½ x9 inches, (*The Chrysler 300-C America's Greatest Performing Car.*)
2—Fact Sheet, No. (C), 2 pages, 8x10 inches. Printed by Ross Roy, Inc. Details differences of Fire Power V-8 with mechanical specifications, performance data chart, list of options and accessories. Dated 1/3/57.
Advertisements
1—*Car Life,* March 1957, page 2
2—*Mechanix Illustrated,* February 1957, page 179
3—*Motor Life,* April 1957, page 11
4—*Motor Trend,* March 1957, page 10
5—*Motor Trend,* April 1957, page 16
6—*Popular Science,* March 1957, page 221
7—*Road & Track,* March 1957, page 4
8—*Road & Track,* April 1957, page 3

This rear shot of Exner's experimental "613" certainly shows off its uniqueness. *(Gil A. Cunningham)*

Arnold Leuth's clean engine is a good example of the 1957 300-C standard type.

Tail-pipe diameters on the dual exhaust system varied with the choice of body style, as well as options. Both the hardtop coupe and convertible came with a standard two-inch-diameter tail pipe, but only the hardtop was available with an optional 2.5-inch exhaust system that included a tail pipe with the larger diameter.

The 1957 hardtop coupe had a distinctive windshield header, with the roof overlapping the glass in a visor-like manner. Unfortunately, this design slightly reduced the top speeds of cars competing at Daytona. The visor look was deleted for 1958 production. *(Linda Clark)*

Leather upholstery with vinyl front seat rear panels was used in the 300-C. The leather had embossed pleats, running side-to-side, for cushion and seat back inserts. The top of the rear seat was designed with a dip in the center. Headliner used on the hardtop was of tan, grained vinyl. On interior door panels, early production cars carried a 300 logo embossed on pressure-sensitive tape. Later cars had the designation embossed into the door panel. Styling was new both inside and outside.

The license plate indentation incorporated a lamp for illumination and there was a round plastic medallion located above it. The circular model badge was held in a spoked double-circle housing. A push-button key lock was positioned directly below the license plate holder.

CHAPTER FIVE
1958 Chrysler 300-D
Respectable Race Car

"The Chrysler 300-D is more than a great new car. . . . it's a unique idea. . . .," stated the copy in the 1958 letter series sales folder, but in reality there were few product changes over the past. The 1957 model run had been a good one for Chrysler products as a whole. Dollar sales of 3.5 billion reflected a 34.5-percent improvement and set an all-time company high. A finned "Forward Look" styling theme, Torsionaire suspension and the excellent TorqueFlite automatic gearbox were the key ingredients for success. The strong market performance of the line—reinforced with a flow of government contracts and heavy investment in plant expansions—resulted in a "leave well enough alone" approach to 1958. This thinking was made official policy on January 10, 1957, when Chrysler President L. L. Colbert publicly confirmed a persistent rumor that the 1960 models would be the next cars that might be accurately called all-new.

For the 300 Letter Car there was only a handful of minor variations this season and an aborted attempt to market an optional electronic fuel-injection (EFI) system. Engines got new air cleaners, a more conventional crankshaft and a less exotic cam. Stylingwise,

The front end of the 300-D was identical to the 300-C, although some late production units may have sported a simplified grille. Dual headlamps were used on all cars since they were now considered legal in every state. *(Jack Wiltse)*

the visored windshield header was deleted from the hardtop coupe, smaller taillights appeared and finish was expanded to a choice of six new standard colors. Interiors were treated to redesigned upholstery and door trim panels, carpet heel pads and more chrome decorations on the dash.

One possible reason for minimal appearance alterations on the 1958 models is related to Virgil Exner's health. In the fall of 1956, at the time when preliminary planning would normally commence, the design chief was struck by a massive heart attack. This almost certainly affected the studio staff, which was not very large to begin with. Although he ultimately recovered, early reports of Exner's condition were grim and the future of the Chrysler 300 styling program seemed unpredictable. The simplest solution was to carry the recently finalized 1957 look over for another year with only the smallest of detail changes.

Although they are virtually identical from the outside, it is not hard to tell the 300-C and the 300-D apart at a car show. The smaller taillights used for the later year are, obviously, one means of differentiation. Forward roof treatment on the hardtop is another. There were also new wheel covers with embossed, red-painted squares at the outer edge. Mounted to the center was a cone-shaped spinner with three fins and a round plastic insert in the middle. This circular medallion was finished in red, white and blue and had the 300 designation in chrome. The most apparent identifier on the 300-D was, however, the large medallion breaking the horizontal spear on the rear flank. It carried the 1958 model name in bold characters.

The 300-D was both unique and historically significant in terms of Letter Car engineering. It was the first and only car in the series available with the EFI power plant option and it was also the last Beautiful Brute to feature the powerful hemi V-8. In addition, some enthusiasts insist that the 390-hp fuel-injected D's were the fastest of all the Fire Power models. Gil Cunningham, a marque expert and also a Chrysler employee, disagrees: "There is no reason to believe it [the EFI option] faster than the 390-hp 'C.'"

The standard Chrysler 300-D engine, in terms of performance, was a far cry from either of these motors or even the base version used a year earlier. In an

The experimental Diablo Dart, of 1957, clearly influenced the Chrysler 300-D's styling treatment. This handsome show car featured high-rising fins and a tasteful trim package that set a theme for production models.

The Dual-Ghia was a limited-production automobile inspired by Chrysler's Firearrow show cars of the early fifties. The design philosophy of Virgil Exner was embodied in the shape of this 1958 edition. It is owned by Packard Industries, Boonton, New Jersey. *(Michael A. Carbonella)*

effort to increase Letter Car business, the company sought a compromise between its original state-of-the-art street racer and the luxury/performance type of machine that some buyers seemed to desire. Thus, the newer motor had an increased horsepower rating combined with a somewhat more civilized demeanor.

The new camshaft, with valve overlap of fifty-five degrees, provided easier starting and a smoother idle, which made a lot of sense from the sales angle. As I found out while photographing cars for this book, the hemis are, sometimes, difficult to start and keep running. In a few cases, it took a good deal of tinkering, cussing and finger-crossing to move a unit to a photogenic locale and several cars were left running while my pictures were snapped. This is an indication of why Chrysler went to the milder cam. At the same time, the company raised the compression ratio of the standard V-8 to 10.0:1. This provided five additional horsepower and looked like a performance improvement on paper.

Of course, this was not an engine for the true enthusiast driver, but the fuel-injected version may have been. Chrysler announced its intentions to release the EFI system as a production option on *some* of its car lines on September 28, 1957. A handful of Chrysler 300's, Dodge D-500's, DeSoto Adventurers and Plymouth Furys were so produced. Actually, Bob Rodger's engineering staff and technicians from the Bendix Corporation's Automotive Electronics Division had been working on it since 1954. In February 1957, a white 300-C coupe with the EFI engine under the hood appeared at Daytona. Once the car hit the sand a lot of testing was done, tremendously straining the drive train. Later, the car was turned over to Vicki Wood to race, but the clutch disintegrated at 120 mph bringing tears to the Detroit grandmother's eyes. (Wood later set a 1957 speed record in a TorqueFlite-equipped C complete with Hi-Way Hi-Fi under the dash.)

Making EFI an assembly line option was simply Chrysler's attempt to compete with fuel-injected offerings from other manufacturers including the Pontiac Bonneville, Corvette, Chevrolet and the surprising Rambler Rebel. It was simply a bolt-on equipment package selling for about $400. Cars that got the option were first fully assembled at the Chrysler Divi-

From the front, the 300-D coupe could be distinguished from a 1957 model only by the design of the windshield header. The 1958 profile was very similar to that of the 300-C. New wheel covers, taillights and model medallions were seen. Minor revisions to the bumpers increased the total length slightly.

Panel alignment for 1957 and 1958 was a point of criticism even when the cars were new. Sometimes adjustments were required for proper fit. Hidden changes to the front door pillar were the additions of a clip at the upper curvature and a sponge-rubber hinge seal at the bottom. This special-order Sandalwood Metallic painted hartop belonging to Jack Wiltse of Michigan has the rarer '57-type headliner.

sion Jefferson plant. They were then driven to the DeSoto factory on Warren Avenue, where the stock carburetors were replaced with the Bendix unit. The first example was built January 20, 1958. According to previously published accounts, thirty-five Chryslers, twelve Dodges, five DeSotos and two Plymouths were assembled with EFI between then and July 15. These figures were tallied by the superintendent in charge of production of these cars at the DeSoto plant. They cannot, however, be substantiated by microfilm records today.

The 300's with fuel injection had special side medallions denoting the installation. Basis for the system was a metering procedure controlled by electric, solenoid-operated valves which ejected fuel ahead of the intake area. Electric fuel pumps and a 40-amp generator were required. The EFI was overly complicated and prone to frequent failure. Few buyers felt that ten extra horsepower were worth $400 and nobody needed the headaches that came with tuning and operating the cars. Chrysler eventually issued a recall and replaced almost all of the Bendix units with stock, dual carburetors. Popular legend has it that just one car escaped the recall and exists today in either Hawaii or Kentucky. According to Gil Cunningham, "Several Chrysler people I have talked to confirm that the one car remaining with EFI was owned by an electronics expert in the Midwest. He apparently sent several letters to Chrysler informing them of its performance." Cunningham also feels that the total output of EFI cars may be as low as half the thirty-five units recorded by the DeSoto production supervisor.

Surprisingly, between the time research on this book began and the date it went to press, news about three other existing EFI-equipped Letter Cars has come to light. One of these cars was advertised for sale in a recent issue of *Brute Force* magazine. Two others have been purchased by Sherwood Kalenberg, of the Walter P. Chrysler Club. In a telephone conversation, Mr. Kalenberg revealed that one car is in complete condition—with the Bendix fuel-injection system intact. It is currently receiving a full professional restoration to show condition.

There were some additional technical changes for 1958. Most were in line with the new emphasis on

Gary Goers received this Ermine White 300-D convertible from his wife Gerri in 1976, prior to an important concours at Anaheim, California. He completed a meticulous ground-up restoration and took Best of Show honors. In 1978, the car was seen in the centerfold of *Motor Trend* magazine. It has since been sold to singer Richard Carpenter who has a collection of Letter Cars. *(Old Cars Weekly)*

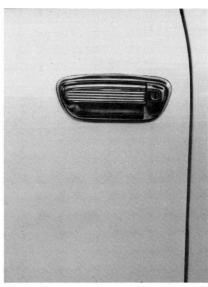

Door handles for the 300-D's were of the same type used in 1957 and had a semi-recessed design. The hand-pull section had four horizontal grooves and the flat keyhole portion was toward the back. *(Berg)*

luxury over total performance. The axle ratio selection was decreased to just five choices. On October 9, 1957, an automatic driver-assist option—called Auto-Pilot—was introduced as a factory accessory and offered for the 300 line, except EFI-equipped cars. Dual headlights, which were introduced in 1957, but not allowed in all states, were now made standard equipment, due to adjustments in the applicable state laws. Another unique option was Captive Air tires, which had an air-filled inner chamber that allowed a car to be driven for some distance on a flat. (A Plymouth test car was once taken from New York City to Boston on only the Captive Air chambers).

Professional racing of Chrysler 300's was on the decline this year, although a move into dry lake racing was precipitated by Norm Thatcher, who took a Letter Car to the Bonneville Salt Flats in Utah. Running in E gas/sedan class competition, Thatcher set a record of 156.387 mph. Brewster Shaw won the Standing Start Mile at Daytona with a stick-shift D coupe. He then did the beach quarter-mile in sixteen seconds at 84 mph. Also appearing at Daytona was a relatively new 300 pilot named Gregg Ziegler. The owner of a hardware store in Elgin, Illinois, Ziegler had brought his own 300-B to Florida in 1957, but this year had the chance to use Dick Dice's 300-D. He got the car up to about 138 mph. In later years, Ziegler would get the chance to set records with an extremely hot, super-special 300-F.

But, I'm jumping the gun. For now suffice it to say that the 300-D had become a different type of "Brute" . . . a type that could be called a more respectable race car.

SPECIFICATIONS
Model number   300-D (Chrysler LC3-S or LC4)
Serial number data   (location) front door hinge post (start) LC4-1001 (end) LC4-1810
Engine number data   (start) 58N3-1002 no additional data
Start production   October 1957 (1 car built October 10, 1957)
Announcement date   December 5, 1957
Advertised dealer prices   (hardtop coupe) $5,173 (convertible) $5,603

Massive bumper "wings" marked the 300-D's profile. The new wheel covers had short, raised "spokes" with the recessed areas painted red. The center of the discs had a red-white-and-blue medallion with chrome model numbers in the center. Surrounding the plastic insert was a chromeplated cone with three fins. The whitewalls on Madeline Berg's rare Chrysler are of a modern design that does not match the factory original type. *(Berg)*

The gas filler door was located in the left rear fender above the side spear. If out of adjustment, paint-chipped edges will result. Fuel-injected cars used a distinct medallion with lettering indicating this special system in place of the normal 300-D type. *(Berg)*

Standard equipment
Special hemi engine, heavy-duty suspension, custom steering wheel, Safety-Cushion dash panel, prismatic rearview mirror, nylon racing-type white sidewall tires, electric clock, directional signals, hand-brake warning signal, power brakes and power steering (automatic transmission only), TorqueFlite transmission, chrome stainless steel wheel covers, windshield washer, undercoating and hood insulation pad, leather upholstery, dual headlamps, Silent-Flite fan drive, rear license plate frame, twin four-barrel carburetion, carpeting.

Options and accessories
(Code) — description — price
(397) — Auto-Pilot — $88.10
(341) — air conditioning, trunk installation — $540.10
(342) — air conditioning, DeLuxe-Dual — $688.60
(443) — Electro Touch-Tune radio and antenna — $124.10
(375) — Bendix electronic fuel injection (EFI) — $400
(37E) — above with air conditioning — price not available
(441) — Music Master radio and antenna — $99.80
(327) — rear shelf radio speaker — $15.10 (hardtop only)
(334) — power radio antenna — $25.80
(401) — manual outside mirror, left-hand — price not available
(403) — manual outside mirror, right-hand — price not available
(402) — remote-control outside mirror, left-hand — $13.80
(405) — vanity mirror — price not available
(394) — rear window defroster — $20.85 (hardtop only)
(345) — Custom Conditionaire heater — $93.30
(347) — Instant Heat heater — $158.45
(304) — power windows — $107.60
(303) — 6-way power seat — $101.90
(467) — Solex tinted glass — $15.90
(469) — shaded backlight — $26.90 (hardtop only)
(481) — rear quarter stone shields — price not available
(399) — Sure-Grip differential — $51.70
(57A-F-H) — 40-amp generators — prices not available
(*) — low back-pressure exhaust system
(381) — front license plate frame — price not available
(*) — manual transmission with manual steering (less power steering, brakes and A/C)

Original paint colors
(factory code) name   Ditzler combination
(AAA)   Raven Black   9000
(HHH)   Aztec Torquoise   42150
(MMM)   Mesa Tan   21447
(OOO)   Tahitian Coral   70779
(PPP)   Matador Red   70791
(XXX)   Ermine White   8131

Interior trim
Standard upholstery color   Tan
Standard upholstery material   Pigskin-like leather with vinyl panels
Seat welting, color/type   Tan vinyl
Instrument panel finish   Sahara Tan
Package shelf   Black vinyl plastic

Convertible top
Color   Black or White
Material (per supplier)   Double-stitch-seam canvas or laminated-seam canvas

Engine (Optional engine figures are for the EFI-equipped V-8)
Type   90-degree Fire Power V-8, overhead valves, cast-iron block
Bore x Stroke   4.00x3.90 inches
Mains   five
Displacement   392 cubic inches
Standard bhp   380@5200 rpm

Taillights for the 300-D were of a new vertical design with shorter lenses. The housings looked similar to the previous type, but were also new and had a more rounded look at the top. The area between the lens and the fender was painted silver. The bumper was constructed of upper bar, lower bar and dust shield, all of which were chromeplated. *(Berg)*

Door panels had beige vinyl upper sections, pearlescent ivory pigskin-like leather Flight Sweep inserts and black-carpeted lower panels. The various divisions were separated with stainless steel moldings. At the lower front corner there was a grouping of five vertical slashes. Power windows were optional. *(Berg)*

The rear license plate was again installed in a "shadow box" recess surrounded by a chrome molding band. The trunk lock was located below it. Mounted directly above was a 300 medallion in a round, chromeplated holder. *(Berg)*

Optional bhp 390@5200 rpm
Standard torque 435@3600 rpm
Optional V-8, torque same as above
Standard compression 10.0:1
Optional V-8, compression same as above
Camshaft drive chain type
Valves
Intake head diameter 2.0 inches
Exhaust head diameter 1.75 inches
Intake stem diameter .372 inch
Exhaust stem diameter .372 inch
Overall length 5.093 inches (intake) 5.05 inches (exhaust)
Cam duration (both V-8's) 276 degrees (intake) 276 degrees (exhaust) 55 degrees (overlap)
Carburetion (standard 1958 V-8 only)
Type (2) Carter WCFB four-barrel
Carburetor models (front) WCFB-2741S (rear) WCFB-2742S
Standard exhaust (hardtop) 2¼-inch diameter at manifold with 35-inch mufflers both sides and 2-inch tail pipe
Standard exhaust system (convertible) 2-inch diameter at manifold narrowing to 1¾ inches. One 15-inch muffler and one 12-inch resonator each side.
Optional exhaust system (hardtop only) 2½-inch diameter front-to-rear
Air cleaners dual (2) dry-element type
Electrical
Original battery type Auto-Lite 12H-70 or Willard HO-12-70, 12-volt negative ground
Original starter Auto-Lite MDT-6003
Original generator Auto-Lite GJC-7013-B
Original generator with EFI option Auto-Lite GGA-6007-C
Original voltage regulator Auto-Lite VRX-6201-A
Transmission
(TORQUEFLITE)
Type three-speed automatic; torque converter with planetary gears
Ratios (low) 2.45:1 (drive) 1.45:1 and 1.0:1
Control location push buttons, left-hand side of dash
(STANDARD TRANSMISSION)
Type three-speed manual
Ratios (low) 2.31:1 (2nd) 1.55:1 (3rd) 1.0:1
Lever location column-mounted w/conventional pattern
Differential hypoid drive type, semi-floating
Axle gear ratios (standard) 3.31:1 (optional at no charge) 2.93:1, 3.15:1, 3.54:1 and 3.73:1
Running gear
Standard power steering Chrysler Constant-Control type of gear-shaft-and-sector design with power piston. Overall ratio 19.38:1 with 3.5 turns lock-to-lock.
Manual steering (w/manual shift only) 3-tooth roller design with symmetrical idler arm linkage
Power brakes Chrysler Total-Contact type. Primary and secondary linings (front and rear) measure 12.6x2.5x13/64 inches. Swept area is 251 square inches. Duct cooling
Chassis and body
Frame type welded, double-channel box section
Body construction welded steel
Wheelbase 126 inches
Front tread 61.2 inches
Rear tread 60.0 inches
Overall length 220.2 inches
Overall height (loaded) 55.2 inches (hardtop) 55.6 inches (convertible)
Overall width 79.6 inches
Ground clearance 5.1 inches
Shipping weight 4,305 pounds (hardtop) 4,475 pounds (convertible)

The 1958 interior was designed to give a sporty, four-place appearance combined with the convenience of six-passenger seating. Rear seat backrests had a dip at the top center and a 300 logo was set into this indentation against the black vinyl package shelf. Two types of headliners were seen in production. Units built early in the year had the same design used in 1957 (now rare), but most cars had a new "molded" type with tan vinyl material and three stainless steel bows which ran side-to-side. *(Berg)*

The dashboard was painted Sahara Tan and the instruments were set into two large circular housings. A 150-mph speedometer was to the left, while the fuel, oil pressure, temperature and amp gauges were clustered in the right-hand circle. Trim plates with a diamond-pattern finish set off the transmission push-button control panel, radio and ashtray, giving a "chromier" look to the dash. This car has a nonoriginal under-dash air conditioner. *(Berg)*

Fuel tank capacity   23 gallons
Suspension
Front, type   torsion-bar with ball joints
Spring rate   170 pounds
Rear, type   7 semi-elliptic leaves
Spring rate   135 pounds
Cooling system
Radiator   full-length water-jacket cooling with thermostatic bypass temperature control
Coolant capacity   25 quarts with heater
Wheels and tires
Tire size   9.00x14
Type   Goodyear Blue Streak, whitewall, tubeless
Whitewall width   2.75 inches
Standard disc wheels   14x6.5
Production totals
Two-door hardtop coupe   618
Two-door convertible   191
Total   809
Remaining 1981*   145 (17.9 percent)
Notes: Total EFI installations recorded at DeSoto factory: 35 units. Survival rate is highest of all Milestone Letter Cars on percentage basis.
*See Chapter Two
Sales literature
1—Folder No. CS-415, 12/57, 8 pages, black & white, 14x11 inches, *Presenting the Chrysler 300 D for 1958.*

Advertisements
1—*The New Yorker* (note: Only ad known, issue date not available)

Front and back seats for the 300-D were upholstered in natural beige cowhide with embossed pleated inserts of pigskin-like pearlescent ivory leather. The overall pattern differed from 1957 in that the pleats now ran front-to-rear instead of side-to-side. The rear of the split backrest was also covered with beige leather and had matching vinyl borders. The prismatic rearview mirror was positioned farther left this year and the radio speaker grille was at the top of the dash. The radio was set into the center of the dash with the heater directly below it. *(Jack Wiltse)*

Horns for the 300-D came from three suppliers—Jubilee, Spartan or Auto-Lite. They were mounted between the radiator and grille. Some cars came with three and others with two. In either case, the horn on the right-hand side was longer. The Auto-Lite type is shown here. *(Berg)*

Innovative features of the 1958 fuel injection system included primary and secondary throttle bodies, fuel pressure regulator, vapor separator and trigger selector. This shot of this clean, rare engine was provided by Gil A. Cunningham.

The air cleaners and fuel line routing on the 300-D's looked distinctive. Pins and latches for the hood, relays for horns and starters, were Number 2 Cadmium-plated to a golden finish as in 1957. Another similarity was that all unpainted linkage components and bolts had silvery, Number 1 Cadmium-plating. *(Berg)*

The engine block, cylinder heads, intake manifold, oil pan, water pump, power steering unit and front engine cover were painted silver. The twin air cleaners and valve covers were finished in gold with distinctive decals applied. Spark plug covers were black, as were the crank and idler pulleys, starter, oil fill cap, generator, voltage regulator and fan. Engine detailing is an important step in preparing a Letter Car for concours judging by either of the two marque clubs. *(Berg)*

CHAPTER SIX
1959 Chrysler 300-E
The Lion-Hearted Letter Car

The 1959 Chrysler 300-E was a good indication of the fact that the Detroit horsepower race had entered a temporary holding pattern. It was the first Letter Car that did not get a boost in power output. Chassis engineering was carried over from the previous season with one big exception: The Fire Power hemi V-8 was gone! In its place was a smoother-running wedge-head engine with an unchanged 380-hp rating. Styling modifications included more glimmer and glitter than any other member of the 300 series would see.

This was a car that stirred up purists when it was new, and still sparks debate among collectors today. It was more of a luxury model than a performance machine, which was part of Chrysler's overall product plan.

The company seemed to have two goals in mind. One was a major investment in research geared to the development of a compact line (the Valiant). But the second was the creation of a car to go head-to-head against the new four-seat Thunderbird. Ford's luxury sports car had no direct competition and was climbing in sales while all other marques except Rambler declined. Chrysler had also had luck in the higher-

The 1959 Chrysler 300-E was introduced in December 1958 for the 1959 model year. Sheet-metal changes over 1958 were minor and the big news was a brand-new 413-cubic-inch, wedge-head engine under the hood. High-performance drive train options were drastically reduced for 1959 and the car had more luxury features than earlier 300 Letter Cars. This Turquoise Gray example is owned by Harry and Linda Ewert of Illinois.

priced bracket with the Imperial. The luxury car represented a full 24.2 percent of divisional volume during 1957 and 21.6 percent the following year. So there was every reason to believe that a smoother, more luxurious Letter Car would prove popular, too.

Styling modifications for the 300-E included an improved grille with narrow horizontal strips, new hubcaps and a generally higher level of trim throughout. The rear end received a totally revised treatment with a smaller deck, shorter taillights, repositioned quarter panel decorations and a massive three-bar bumper incorporating a beauty panel. The major design theme was still one of clean lines, restrained application of ornamentation and aircraft motifs rendered with a European flavor. *Motor Trend* selected the 300-E as the season's "best-looking hardtop," specifically noting that the choice was predicated on the coupe's clean look. After photographing the cars for this book, I agree that the fifth Letter Car was an attractive and beautiful machine in an age of styling excesses. However, it does have quite a bit more trim than the previous models carried.

A clear reflection of the Chrysler 300's new status for 1959 shows up with a glance at the regular equipment features and changes in the option list. A tilt-type prismatic rearview mirror which had an electric eye that made angle adjustments whenever the backlight became too bright, bright metal sill moldings, Living Leather upholstery and individually swiveling seats were standard. New accessories included an automatic headlamp-beam changer, Mirromatic mirror and True-Level air-suspension system. But, the big difference here was in option *deletions*. No longer available were higher horsepower engines with exotic camshafts and solid lifters. Gone as well were choices of different transmissions or rear ends. All 300-E's came with the push-button TorqueFlite gearbox and had either a 3.31:1 or 2.93:1 axle. Power steering and brakes were mandatory.

Chrysler claimed that the new, hundred-pound-lighter Golden Lion engine could out-perform the hemi. It displaced 413 cubic inches and featured many of the usual goodies such as twin four-barrel carburetors with staged throttle linkage, a relatively

The front of the 300-E was distinguished by a revised horizontal-motif bar grille, changes to ornamentation and red-orange finish for the secondary radiator grille bars and brake scoop grilles. Some factory literature referred to this contrasting grille paint as a "two-tone" combination. *(Ewert)*

The original tires were 9.00x14 Goodyear Blue Streaks. According to some sources the whitewall width was 2 5/8 inches. However, there are other sources which indicate a 2¾-inch white sidewall was used. The hubcaps have red-white-and-blue medallions, with chrome 300 numerals. These badges are set into an anodized aluminum, gear-shaped hubcap. The wheel covers are of chrome stainless steel, with the contrast ring finished in flat black. *(Ewert)*

high-lift cam, heavy-duty valve springs and dampeners, dual breaker points, low-restriction air cleaners, special intake manifold and a low back-pressure exhaust system. Still, enthusiasts had their doubts about the wedge's acceleration potential, since it didn't have quite as deep a rumble as the Fire Power V-8.

The arguments (which continue to this day) began when *Motor Trend's* Bill Callahan reported, "The 1959 'E' series moves even faster than the 1958 'D' series." He went on to explain that his words were backed with figures recording a 1.4-second-faster 0-60 time for the new car, over the old. This, of course, does not explain the new Letter Car's slower runs at Daytona Speed Weeks. Harry Ewert—the owner of the car pictured in this chapter—believes it was the lack of performance options that made the E slower than *some* of the earlier models. According to Ewert, *Motor Trend* tested 300's with the base engine each year and, in this form, found the 413-powered Letter Car was quicker. It could not, however, match the 390-hp jobs. Some owners commented that the steering was too light.

In reality, the new 300 was not nearly the "slug" that some articles made it out to be. *Speed Age* magazine gave a white coupe the workout in March 1959 and discovered that it could move from 0-60 in 8.3 seconds and reach the century mark, from the same starting point, in 22.5 ticks of the clock. This car also got 12.4 mpg during its exercise. A similar report in the August 1959 issue of *Sports Car Illustrated* noted a quarter-mile time of 17.2 seconds with a terminal speed of 92 mph. But, perhaps the E was summed up best by *Consumer Reports* which stated, "Certainly, the 300-E is docile enough in its ordinary road behavior so that you can forget about its sports-car potential. It may ride a little hard for a *Chrysler* hardtop, but it's *almost* a car that's suitable for use anywhere in the U.S."

The 300 series sales booklet this season alluded to "lessons learned at Spa, Le Mans and Watkins Glen," but the E did not have much of its own racing history by the end of the year. At Daytona, on Sunday, February 15, 1959, Gregg Ziegler attacked the beach again. Speeding south, one-way over the sand, he reached 120.481 mph. That was enough to place driver and

The round, red-white-and-blue plastic identification medallion was also seen on the passenger side of the car. It was circled with a bright metal ring and had short extensions at each side which overlapped the long spear, running toward back of car, and the short spear between the medallion and door opening. *(Ewert)*

The three-quarter front view showed clearly the gently rising fin atop the hood center and the new 300 letter designation found only on the driver's-side front lower corner of the hood. Parking lamp lenses were white plastic. This grille is the N.O.S. stand-in used for car shows and special events. *(Ewert)*

mount in the exclusive Century Club once more; but three Chevys, two Pontiacs and one Corvette did better. In quarter-mile acceleration runs, the Pontiacs driven by Bob Pemberton and former Chrysler pilot Vicki Wood left only third place for the 300-E, which terminated its run at 87.08 mph. And, the one car running in the 500-mile stock car race went out with a blown engine.

Another thing that Chrysler blew was its hopes of tapping the Thunderbird market with its new type of Letter Car. Total production for the series was 119 units *less* than in 1958. The 300-E failed to lure the type of buyer that wanted a distinctively styled sports/luxury machine, because it did not look vastly different from other Chrysler products. At the same time, the kind of customers who had preferred the earlier editions in the series, found the new one unsuited to their tastes. Gregg Ziegler, for example, later referred to the car which he purchased to take to Daytona as a "kind of modified New Yorker." The company would soon rectify its mistake, but only after a painful lesson in 1959.

SPECIFICATIONS
Model number   300-E (Chrysler MC3-E)
Serial number data   (location) front door hinge post (start) M591100001 (end) M591100690
Engine number data   (location) not available (start) not available (end) not available
Body numbers (new data)   (hardtop) 592 (convertible) 595
Start production date   not available
Announcement date   December 1958
Advertised dealer price   (hardtop) $5,318.50 (convertible) $5,748.50
Standard equipment
Special wedge engine, heavy-duty suspension, custom steering wheel, Safety-Cushion dash panel, prismatic rearview mirror, nylon high-performance white sidewall tires, electric clock, directional signals, handbrake warning signal, power brakes, power steering, TorqueFlite transmission, chrome stainless steel wheel covers, windshield washer, undercoating, hood insulation pad, leather upholstery, dual headlamps, Silent-Flite fan drive (limits fan speed to 2500 rpm), rear license plate frame, swivel seats, low back-pressure exhaust system (hardtop only).

Rear styling treatment was totally new and a bit glittery for a Letter Car. It featured bigger bumpers, smaller taillight lenses and hood-like fin tip extensions. A decorative beauty panel was set into the bumper opening with backup lights between bumper bars. Numerical model designation decorated passenger-side deck lid corner. *(Ewert)*

Vinyl-covered armrest was near center of door. Bottom was trimmed with a horizontal band of black cut-pile carpet. Stainless steel moldings separated the different upholstery materials. Vent windows were push-out type with manual flip-up latches. Controls for operation of all four windows were mounted on driver's door. *(Ewert)*

Trunk ornamentation included company name in chrome letters with black inside trim and round E badge of red, white and blue plastic. Housing for badge had recessed areas finished in black. Design of the decorative beauty panel featured stacked rows of bright-finished raised ribs. *(Ewert)*

Options and accessories
(Code)—description—price
(397)—Auto-Pilot—$86.10
(341)—air-conditioning, unit w/heater —$10.40
(342)—air-conditioning, DeLuxe-Dual w/Instant-Heat heater—$714.25
(443)—Electro Touch-Tune radio and antenna—$124.10
(441)—Music Master radio—$99.80
(327)—rear shelf radio speaker (hardtop only)—$17.30
(334)—power antenna—$25.90
(401)—manual-control rearview mirror, right-hand—$6.50
(403)—manual-control rearview mirror, left-hand—$6.50
(402)—outside remote-control rearview mirror, left-hand—$18.00
(405)—Mirromatic mirror—$17.70
(391)—automatic headlamp-beam changer—$50.30
(398)—rear window defogger (hardtop only)—$20.85
(345)—push-button Custom Conditionaire heater—$136.30
(347)—Instant Heat heater—$136.30
(304)—power window lifts—$107.60
(362)—Six-Way power swivel seat—$101.90
(462)—Solex tinted glass including large, shaded backlight—$70.00
(461)—large, shaded backlight only— $26.90
(396)—True-Level Torsionaire suspension—$144.90
(399)—Sure-Grip differential—$51.70
(545)—extra-heavy-duty 40-amp generator—$71.60
(381)—front license plate frame—price not available

Original paint colors

(factory code)	name	Ditzler combination
(AAA)	Formal Black	9000
(KKK)	Turquoise Gray	42263
(RRR)	Radiant Red	70791
(WWW)	Cameo Tan	21551
(XXX)	Ivory White	8131
(ZZZ)	Copper Spice	21550

Interior trim
Standard upholstery color   Tan
Standard upholstery material   Pearlescent tan pleated Living Leather, smooth-grained natural-beige cowhide, pleated and smooth-grained tan vinyl.
Seat welting, color/type   Tan vinyl
Dashboard color   Sahara Tan
Package shelf   Black vinyl

Convertible top
Color   Black or White
Material (per supplier)   Double-stitched or laminated-seam canvas

Notes:
Code 888 on vehicle data plate indicates special-order upholstery combination
Code 999 on vehicle data plate indicates special paint color combination

Engine
Type   90-degree Golden Lion V-8, overhead valves, cast-iron block
Bore x Stroke   4.18x3.75 inches
Mains   five
Displacement   413 cubic inches
Standard bhp   380@5000 rpm
Optional bhp   none
Standard torque   450@3600 rpm
Optional torque   none
Standard compression   10.1:1
Optional compression   none
Camshaft drive   chain type

Valves
Intake head diameter   2.08 inches
Exhaust head diameter   1.74 inches
Intake stem diameter   .372 inch
Exhaust stem diameter   .372 inch
Cam duration   260 (intake) 260 (exhaust) 42 (overlap) degrees

Carburetion
Type   (2) Carter AFB four-barrel
Carburetor models   (front) AFB-2798S (rear) AFB-2799S

The front seats were bucket-like, individual units separated by a folding armrest and an upholstered center cushion. Tan vinyl welting protected the seams. The custom steering wheel was standard with a two-spoke design having padding across the deep-dished spokes and a chrome horn ring at bottom. *(Ewert)*

The instrument board section of dash was dominated by large round dials with 150-mph speedometer to the left—engine monitoring gauges to the right. Heater controls and ignition switch were to the right of steering column, with light control, wiper control and cigar lighter button on the south-paw side. Push-button shift control was at the extreme left of dash. *(Ewert)*

Standard exhaust (hardtop) dual exhausts, 2¼-inch diameter, (1) 34½-inch muffler on each side (convertible) dual-exhausts, 2-inch diameter, (1) 15-inch front muffler and (1) 12-inch rear resonator on each side
Air cleaners   Dual paper type

Electrical
Original battery type   Willard HO-12-70 or Auto-Lite 12-H-70, 12-volt, negative ground
Original starter   Auto-Lite MDT-6002
Original generator   Auto-Lite GJM-8001A
Original voltage regulator   Auto-Lite VRX-6301-A

Transmission
(TORQUEFLITE)
Type   three-speed automatic; torque converter with planetary gears
Ratios   (low) 2.45:1 (drive) 1.45:1 and 1.0:1
Control location   push buttons, left side of dash

Running gear
Differential   hypoid type, semi-floating
Axle gear ratios   (standard) 3.31:1 (optional) 2.93:1
Standard power steering   Chrysler Constant-Control type, gearshaft-and-sector design with power piston. Overall ratio 19.38:1, 3.5 turns lock-to-lock.
Standard power brakes   Chrysler Total-Contact type. Primary and secondary linings (front and rear) measure 12.6x2.5x13/64 inches. Swept area is 251 square inches. Duct cooling.

Chassis and body
Frame type   welded double-channel box section
Body construction   welded steel
Wheelbase   126 inches
Front tread   61.2 inches
Rear tread   60.0 inches
Overall length   220.2 inches
Overall height (loaded)   55.3 inches (hardtop) 55.7 inches (convertible)
Overall width   79.5 inches
Ground clearance   6 inches
Shipping weight   4,290 pounds (hardtop) 4,350 pounds (convertible)
Fuel tank capacity   23 gallons

Suspension
Front, type   torsion-bar with ball joints
Spring rate   170 pounds
Rear, type   7 longitudinal leaf springs
Spring rate   136 pounds

Wheels and tires
Tire size   9.00x14
Type   Goodyear Blue Streak, whitewall, tubeless
Whitewall width   2-5/8 or 2-3/4 inches

Standard disc wheels   14x6.5
Production totals
Two-door hardtop coupe   550
Two-door convertible   140
Total   690
Remaining 1981*   73 (10.5 percent)
*See Chapter Two

Sales literature
1—Folder No. CS-427 (undated), 8 pages, black & white, 12x12½ inches, 'E' Chrysler 300 E the International Classic That's Made in America.
2—Sheet No. E-7, Oct. 1958, 1 page, both sides, printed by Ross Roy, Inc. Distinguishing features, colors, equipment, specifications, measurement and prices.

Advertisements
1—*National Geographic*, July 1959
2—*Life* (issue unknown)
3—*Time* (issue unknown)

Top center of dash housed prismatic rearview mirror, flip-up air ducts (on cars with air conditioning) and radio speaker grille. The radio and ashtray were on the face of the dash, immediately below. Ribbed heater doors below dash, at transmission hump, were vacuum-controlled. The dash pad was black; painted sections, tan; and trim was of aluminum with a simulated engine-turned pattern. Colors on Code 888 cars could vary. *(Ewert)*

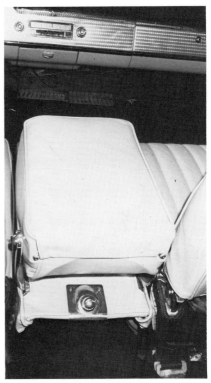

Rear face of front seat dividing cushion housed a cigar lighter for back seat drivers, while rear face of regular seat cushions was carpeted in black cut-pile. Seat belts were an available, dealer-installed option. *(Ewert)*

The 300-E interior was handsome and well-planned. The wide doors on Chrysler products of this era were often praised by Ford and GM owners who had the opportunity to ride in MoPar models. The doorsill moldings had long, horizontal ribs embossed into them with a trapezoidal section stamped with repeating circular indentations and the 300 designation spelled out in raised numerals. *(Ewert)*

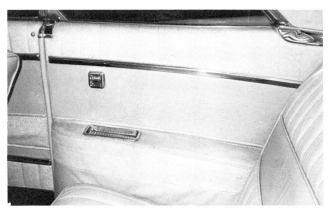

Rear compartment sidewall panels on the 300-E hardtop looked like this. On convertibles, the general appearance was similar, but the tan armrest cover was eliminated due to the intrusion of the folding-top mechanism. Rear seats featured patterns, colors and materials that matched the fronts, including Golden Lion logo embossment on horizontal seat back band. *(Ewert)*

Stainless steel window garnish moldings and headliner bows brightened the cockpit of the 300-E hardtop. The coupe had courtesy lights mounted just forward of the top-front corner of the rear quarter windows. The headliner was covered with pebble-grained vinyl matched to the color of the seats. The rear window package shelf was tan grained-vinyl, tan woven cane or tan-colored panel board, depending on where and when car was assembled. *(Ewert)*

The 413-cubic-inch block can be identified by the designation "M413_ _ _ _" stamped on the boss behind the distributor. A Willard or Auto-Lite battery with six water caps was used. Firewall and inner fender skirts were done in body color, but the hood hinge springs were usually in black. The dual low-restriction air cleaners were painted gold and had a decal (reproductions available) on the side. Windshield washers were standard equipment and the bag for the system was found on the passenger-side inner fender. *(Ewert)*

The valve covers were painted black and had metallic-finish stickers at the middle. The sticker had a checkered-flag design with the 300-E designation spelled out in bold red numbers and letters. There were two blue border lines (reproductions available). Drive belts and radiator hoses were black rubber, and squeeze-ring-type hose clamps should be used. The "wedge" engine developed 380 hp at 5000 rpm, which was the same as the carbureted 1958 hemi. *(Ewert)*

CHAPTER SEVEN
1960 Chrysler 300-F
Sixth of a Celebrated Breed

In January 1959, officials of Chrysler, the Piasecki Aircraft Company and Curtiss-Wright announced that the three firms were planning to develop a flying automobile. This had nothing to do with the letter series, of course, but the 1960 Chrysler 300-F turned out to be a machine that could really "fly."

Hailed as the "Sixth of a Celebrated Breed," the 300-F reflected a revived emphasis on high-performance and was, probably, a part of the last-ditch effort to salvage a falling management regime. Tex Colbert's term as company president was coming to an end, but it was hoped that giving the Letter Car some extra punch could forestall impending doom. Racing was back in the picture, too. With national television coverage planned for the Daytona Beach Safety and Performance Trials, Chrysler hoped to reap valuable publicity by projecting a new performance image.

The 1960 edition of *Ward's Automotive Yearbook* announced the mid-year introduction of the 300-F, referring to the latest product as a "sports car" with a standard 375-hp ram-induction engine. According to the editors, "Of very limited availability was an op-

George Cone and Bruce Hoover, of Illinois, are the co-owners of this white, four-speed 1960 Chrysler 300-F convertible which has the rare Pont-à-Mousson transmission. According to documents in the owners' possession, it was purchased in 1960 from Edwards Motor Company of Milwaukee, Wisconsin, for $6,200. It has had a complete restoration—and captured first place at the Allstate Concours in 1980 and 1981.

tional 400-hp version with an imported, French Pont-à-Mousson manual four-speed gearbox."

Styling for 1960 was not dramatically changed although the product looked fresh from every angle. The basic ingredients were the same finny, aircraft-inspired touches seen earlier, but a zestier, zoomier character was evoked by a number of subtle alterations. Several of these, including a gaping grille with crossed division bars, four individual bucket seats, a full-length center console and a dummy continental tire impression on the deck, were adapted from the "613" styling prototype. Body panels were more highly sculptured and had a crisply tailored, angular look. The rear styling treatment was most highly revised and incorporated sculptured rear quarter panels, reshaped fins with a sexy outward flare and a flat, swept-back bevel for the deck lid latch panel.

There was a great deal of modernization, reflected in the handling of small design details. The blacked-out grille, louvered hood and console-mounted tachometer would soon become identifying trademarks of most sixties-style Detroit "muscle cars." While the C-300 had been America's original high-performance street machine in 1955, the 300-F was the first of the super-hot factory stockers to come with all the trim and trappings. This Beautiful Brute served as a model for the most exciting era of U.S. automotive history.

Engineering improvements included unitized body construction, a freer-breathing base power plant, the special high-output engine option and expanded rear axle ratio availability. Both motors were based on the 413-cubic-inch-displacement Golden Lion V-8 with 10.1:1 compression wedge-design cylinder heads. Equipment common to both included higher-output camshafts, heavy-duty valve springs and dampeners, low back-pressure exhaust, dual-breaker distributor, low-restriction air cleaner, special spark plugs and Silent-Flite fan drive.

An important advance on both motors was the new ram-tuned induction setup. Chrysler had experimented with the concept of ram-induction as early as 1953 in preparing the Kurtis-bodied Indy car. It had featured Hilborn fuel-injection with eight, long ram stacks snaking out above the manifold. By clearly demonstrating the advantages of ram-induction, the race car had sealed its own fate and was outlawed

The French Facel Vega was introduced in 1954 and featured a Chrysler Fire Power hemi V-8 and a Pont-à-Mousson four-speed manual gearbox. A similar transmission was later used for a limited number of 300-F's. Shown is a 1957 model. *(Newport Automobile Museum)*

prior to the running of the 1953 Memorial Day classic. However, the late fifties saw continued experimentation with ram-induction by Chrysler, while Mercedes-Benz and Pontiac (on the Bonneville) toyed with the concept for production cars. However, ram-tuning, up to this point, had always been used in conjunction with fuel injection.

A problem in adapting ram-induction to carbureted production cars was the fact that extremely long ram stacks were required. Such stacks could be easily accommodated on racing machinery, but were difficult to squeeze below the hood of a show room model. Chrysler overcame this by criss-crossing thirty-inch-long tubes above the top of the engine and by reversing the normal positions of the carburetors and ram tubes. The carbs were mounted at the outboard end of the tubing, providing a more practical setup for an assembly line automobile. The resulting boost in air flow was comparable to the effect of a supercharger. The ram-inducted 413-cid motor developed maximum torque of 495 pounds-feet, a ten-percent increase over 1959.

The Chrysler technicians also found a drawback in the thirty-inch-long ram-tubes, in that they were less effective at higher engine rotational speeds. The "long tubes" worked great at around 3000 rpm, but actually hurt performance in the 4000 rpm range. This secondary problem was resolved on the optional Letter Car engine by removing a section of the inner tube walls to give the same effect as halving the length of each tube. The resulting "short tubes" produced slightly lower torque (465 pounds-feet) at a very high rpm level (3600 rpm) and proved extremely effective in racing, as they substantially boosted top speed.

The 400-hp "short tube" engine also had a longer-duration camshaft, bigger valves and solid valve lifters. It was essentially a competition engine and was provided, almost exclusively, to buyers with professional racing in mind. The Pont-à-Mousson gearbox was *exclusive* and *mandatory* with this motor and the package price for the setup was steep. Gregg Ziegler purchased such a car after receiving a phone call from Chrysler's Bert Beauchamp. Would Ziegler

This factory-new Formal Black 300-F hardtop is the car that Gregg Ziegler drove to a Daytona Beach speed record in 1960. The 400-hp four-speed-equipped car covered the Daytona Flying Mile at a 144.927-mph average. Bob McAtee, of Illinois, owns the car today which has just over 11,000 original miles on it.

Manual or remote-control outside rear-view mirrors were still an option for the Letter Car. A louvered hood was a new touch. Cars equipped with radios had the antenna mounted to the right-hand fender top. *(McAtee)*

be willing to spend $800 extra on his F to get 400 hp and a four-speed? he asked.

This imported gearbox was adapted from the Facel Vega, an elegant grand touring car that had first appeared, for the French luxury-class market, in 1954. Facel Vega features included the Chrysler Fire Power V-8 and the aluminum Pont-à-Mousson four-speed. Thus, the transmission was little more than a bolt-in item for the Letter Car, with the high price covering some extensive floor pan modifications which were required for installation on the 300-F, plus the additional engine hardware. The high-performance package was never intended for more than a handful of cars, and surviving examples are now considered very collectible, valuable and rare.

This rarity factor is extremely important to marque collectors as it can double—or even triple—the value of an F. It it definitely known that seven such cars were built. Six of these were hardtops entered at Daytona Beach by Larry Rouse, Warren Koechling, Danny Eames, Harry Faubel, Jr., Brewster Shaw and Gregg Ziegler. Two (owned by Bob McAtee and Bob Wilkins) are known to exist today. The seventh car was a modified (supercharged) hardtop owned by Andy and Joe Granatelli, which was raced at the Bonneville Salt Flats. There are persistent rumors that this car is located in Texas today.

Several other cars with the 400-hp/four-speed package have been heard of. In January 1979, marque enthusiasts Bruce Hoover and George Cone found and purchased a convertible with the special drive train. Photostatic documentation supplied to the author shows the ragtop was originally delivered to George Kuehm of Milwaukee, Wisconsin, on April 25, 1960. Kuehm, apparently, had some association with Carl Kiekhaefer, which enabled him to order this unusual car. Other reports indicate that two additional examples are located in California and Colorado today. A number of authoritative articles indicate that as many as fifteen of the cars may have been produced, but there is debate over this statistic which cannot be verified, despite detailed research by many experts. There is no doubt, however, that very few of

A recent restoration, this Formal Black 300-F convertible belongs to Bruce Hoover. It has the standard 375-hp engine and TorqueFlite transmission, which most of the cars used this year. Ram-induction is featured on this one of under 300 ragtops produced in 1960.

The front view of the 300-F was characterized by a new grille, bent-shaped bumper and more angular lines. The radiator grille insert was of egg-crate design and had flat black finish to set off the chromeplated cross-bars. *(McAtee)*

these unusually optioned Letter Cars were ever assembled.

Chrysler's reentry into racing during 1960 was not an out-and-out violation of the AMA resolution, which stopped manufacturers from openly backing or advertising competition activities, but did not ban the *production* of high-performance cars. When other manufacturers such as Ford, Pontiac and Chevy sidestepped the ruling with "under-the-counter" aid to racers, Chrysler was left in the cold. But, by late 1959, advertisements for Dodge and Plymouth became thinly disguised boasts of performance. For 1960, the Dodge D-500 Ram Charger V-8's and SonoRamic Plymouths appeared, along with the ultimate expression of the ram-tuned MoPar embodied in the new 300-F.

There was no disputing the fact that such cars were built for racing, but somewhat surprising was a blatant Chrysler effort to capture headlines and top billing in various media, including television. The idea was to sweep Daytona again with millions of people watching.

The story of the Chrysler 300's return to Daytona Beach was suitably summarized by sports editor Bernard Kahn's article in the Daytona Beach *News-Journal*. It began, "A retail hardware store owner from Elgin, Illinois, yesterday bettered a long standing NASCAR Flying Mile record by averaging 144.927 mph for the two way run over the beach straightaway." The old record had been set by Tim Flock's 300-B in 1956. The new record belonged to Gregg Ziegler, driving a four-speed-equipped, 400-hp 300-F. The other Letter Cars were in the statistics, too, with two-way averages as follows: Larry Rouse, 141.509 mph; Warren Koechling, 142.687; Danny Eames, 143.027; Harry Faubel, 143.198; and Brewster Shaw, 143.369. Shaw also set a Standing Start Mile record of 88.235 mph.

Later in the year, Andy Granatelli entered his 300-F, equipped with a supercharger, in the experimental class at Bonneville and set an average speed record of 184.049 mph with an amazing one-way run of 189.990 mph. Although the 300-F's never made Walter Cronkite's television coverage of Daytona (he focused on Chrysler's Hyper-Pack Valiants instead), it was a good racing year.

The rear quarter panels were decorated with a gently sloping spear and a large, circular medallion. The fins had a boomerang-style notch at the rear, which housed a similarly-shaped taillight lens. *(McAtee)*

The forward profile of the sixth Letter Car was dominated by flat, horizontal lines, sculptured flanks and massive bumper wing tips that wrapped around the corner of the body. Wheel openings were high and wide to show off the beautiful wheel covers. *(McAtee)*

SPECIFICATIONS
Model number 300-F (Chrysler PC3-H)
Serial number data (start) 8403108193
(end) 8403161678
Engine number data not applicable
Start production date December 20, 1959
Announcement date January 15, 1960
Advertised dealer price (hardtop coupe) $5,411 (convertible) $5,814

Standard Equipment
Special wedge engine, heavy-duty suspension, custom steering wheel, Safety-Cushion dash panel, prismatic rearview mirror, nylon high-performance white sidewall tires, electric clock, directional signals, handbrake warning signal, power brakes, power steering, TorqueFlite transmission, chrome stainless steel wheel covers, windshield washer, undercoating, hood insulation pad, leather upholstery, dual headlamps, Silent-Flite fan drive (limits fan speed to 2500 rpm), swivel seats, low back-pressure exhaust system, power windows, tachometer, Flight Sweep deck lid, center armrests (front and rear).

Options and accessories
(Code)—description—price
(311)—air conditioning, unit w/heater—$510.40
(313)—air conditioning, ordered with accessory group #309—$408.50
(423)—Golden-Touch-Tune radio—$124.10
(422)—Golden-Tone radio—$99.80
(427)—Golden-Tone radio with rear speaker (hardtop only)—$117.10
(428)—Golden-Touch-Tune radio w/rear speaker (hardtop only)—$141.40
(324)—left-hand outside mirror—$6.50
(383)—remote-control left-hand outside mirror—$18.00
(375)—automatic headlight-beam changer—$44.40
(339)—rear window defogger (hardtop only)—$20.85
(315)—push-button Custom Conditionaire heater—$101.90
(293)—6-Way power swivel seat—$101.90
(441)—Solex tinted glass—$43.10
(442)—Solex tinted glass including large shaded rear window—$31.20
(379)—Sure-Grip differential—$51.70
(363)—accessory package A: door edge protectors, vanity mirror and license plate frames—$10.50
(607)—antifreeze—$5.90
(298)—vacuum door-locking system—$36.60
(535)—extra-heavy 40-amp generator—$41.40
(309)—Custom accessory group: includes Custom-Conditionaire heater, Golden-Touch-Tune radio, rear seat speaker, power antenna and rear window defroster—$290.05
(***)—400-hp V-8 with four-speed manual transmission—$800.00
Note: It has been noted that Chrysler lost money on every 400-hp/4-speed car built.

Original paint colors
(factory code)	name	Ditzler combination
(BB-1)	Formal Black	9000
(PP-1)	Toreador Red	71003
(WW-1)	Alaskan White	8218
(ZZ-1)	Terra Cotta	71053

Interior trim
Standard upholstery color Tan
Standard upholstery material Smooth-grained natural tan (beige) cowhide, basket-weave textured leather, black cut-pile carpets.
Seat welting, color/type Tan vinyl
Dashboard color Saddle Beige gloss (lower); Black Frost metallic gloss (upper)

Convertible top
Color Black, White and Terra Cotta
Material Canvas-type fabric
Notes:
Code 888 on vehicle data plate indicates special-order upholstery combination
Code 999 on vehicle data plate indicates special-order paint color combination

The tailfins now flared out from the body at the upper rear area. This resulted in shadows and reflections that created bold character lines, which seemed to move in a number of directions at once. They added excitement to the design. *(Cone* and *Hoover)*

The side trim was really a three-piece affair that had a short spear ahead of the red-white-and-blue medallion and a longer rear spear that now continued around the back of the body. Plated rear fender shields were an option. *(Cone* and *Hoover)*

Engine
Type   90-degree Golden Lion V-8, overhead valves, cast-iron block
Bore x Stroke   4.18x3.75 inches
Mains   five
Displacement   413 cubic inches
Standard bhp   375@5000 rpm
Optional bhp   400@5200 rpm
Standard torque   495@2800 rpm
Optional torque   465@3600 rpm
Standard compression   10.1:1
Optional compression   10.1:1
Camshaft drive   chain type
Valves
Intake head diameter (all V-8)   2.08 inches
Exhaust head diameter (std. V-8)   1.60 inches
Exhaust head diameter (opt. V-8)   1.74 inches
Cam duration (std. V-8)   268 (intake) 268 (exhaust) 48 (overlap) in degrees
Cam duration (opt. V-8)   284 (intake) 284 (exhaust) 55 (overlap) in degrees
Type lifters (std. V-8)   hydraulic
Type lifters (opt. V-8)   solid
Carburetion
Type (std. V-8)   (2) Carter AFB four-barrel w/30-inch rams
Type (opt. V-8)   (2) Carter AFB four-barrel w/15-inch rams
Carburetor models (std. V-8)   (front) AFB-2903S (rear) AFB-2903S
Carburetor models (opt. V-8)   (front) AFB-3084S (rear) AFB-3084S
Standard exhaust   dual system, low-restriction type
Optional exhaust (400-hp V-8)   dual system, 2½-inch diameter, low-restriction type
Electrical
Original battery type   Willard H-12-70 or Auto-Lite 12-H-70, 12-volt, negative
Original starter   Auto-Lite MDT-6002
Original starter   Auto-Lite GJM-8001A
Original voltage regulator   Auto-Lite VRX-6301-A
Transmission
(TORQUEFLITE)
Type   three-speed automatic; torque converter w/planet gears
Ratios   (low) 2.45:1 (drive) 1.45:1 and 1.0:1
Control location   push-button, left side of dash
(PONT-À-MOUSSON)
Type   four-speed manual, aluminum case
Ratios   (first) 3.35:1 (second) 1.96:1 (third) 1.36:1 (fourth) 1.00:1
Control location   floor lever
Running gear
Differential   hypoid type, semi-floating
Axle gear ratios   (standard) 3.31:1 (optional) 2.93:1, 3.15:1, 3.23:1, 3.54:1 and 3.73:1
Standard power steering   Chrysler Constant-Control type, gearshaft-and-sector design with power piston. Overall ratio 19.38:1, 3.5 turns lock-to-lock
Standard power brakes   Chrysler Total-Contact type. Primary and secondary linings (front and rear) measure 12.6x2.5x13/64 inches. Swept area is 251 square inches.
Chassis and body
Frame type   Bolted-on front stub frame
Body construction   welded steel, unitized body
Wheelbase   126 inches
Front tread   61.2 inches
Rear tread   60.0 inches
Overall length   219.6 inches
Overall height (loaded)   55.3 inches (hardtop) 55.7 inches (convertible)
Overall width   79.5 inches
Ground clearance   5.8 inches
Shipping weight   4,270 pounds (hardtop) 4,310 pounds (convertible)
Suspension
Front, type   Torsion-bar with ball joints

The 300-F had a large cargo area inside the trunk with the spare housed in the left rear fender region and mounted horizontally. The key lock was positioned in the top contour of the latch lid panel. *(McAtee)*

The trunk latch panel had a beveled look and inward slant at the bottom. It housed a backup lamp at each corner and a "shadow box" type of license plate recess housing. The bumper had a gently bowed look with two small bumperettes. *(Cone* and *Hoover)*

The spare tire impression embossed on the deck had a wide stainless steel trim ring and the standard hubcap and wheel in the center. A small drain hole was incorporated at the six o'clock position. *(Hoover)*

Spring rate   175 pounds
Rear, type   7 longitudinal leaf springs
Spring rate   136 pounds
Wheels and tires
Tire size   9.00x14
Type   Goodyear Blue Streak, whitewall, tubeless
Whitewall width   2½ inches
Standard disc wheels   14x6.5K
Production totals
Two-door hardtop sport coupe   964
Two-door convertible   248
Total   1,212
Remaining 1981*   196 (16.1 percent)
*See Chapter Two
Sales literature
1—Catalog No. CS-349, 2/60, 12 pages, black & white, 13 x 10½ inches. *An Exciting Story for Those Who Appreciate Greatness in an Automobile: Chrysler 300-F.*
2—Sheet No. (C), Jan. 1960, 3 pages, printed by Ross Roy, Inc. Equipment, dimensions, specifications and V-8 data.
Advertisements
1—*Car Life,* September 1960, page 8
2—*Motor Trend,* May 1960, page 2
3—*Motor Trend,* June 1960, page 2
4—*Motor Trend,* July 1960, page 8
5—*Sports Car Illustrated,* August 1960, two-page
6—*Sports Car Illustrated,* June 1960, Back Cover

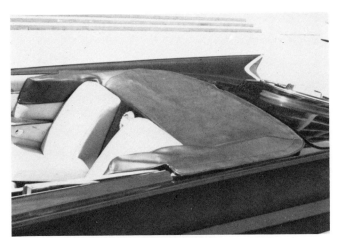

The convertible had a tight-fitting boot with concealed snaps. It covered the top of the rear interior side panels. Standard boot colors were black and red, but at least one car with specially ordered Polar Blue paint had a matching blue boot. The well under the boot was done in black vinyl. *(Hoover)*

The rear of the seat backrests were decorated with wide, bright metal edge moldings, trapezoid-shaped black carpet panels and there was similar carpeting on the rear face of the seat cushion. *(McAtee)*

The interior door panels were finished in a combination of tan vinyl, black carpeting and brushed-aluminum insert material. An integral armrest was a standard feature and housed the door hardware. *(McAtee)*

Most hardtops had smooth-grained beige vinyl headliners. Some late cars came with a perforated white vinyl type. There were three stainless steel bows, courtesy lamps and a handy coat hook. *(McAtee)*

Bob McAtee demonstrates how the folding front armrest on the 1960 Letter Car lifted to reveal a storage compartment. The front seats were upholstered in beige (tan) leather and had a basket-weave-pattern insert and wide horizontal pleats.

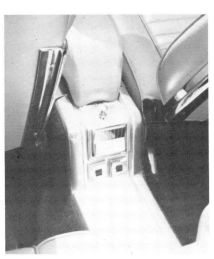

A full-length console separated the four bucket seats. A shorter rear console housed convenience items like an ashtray, cigarette lighter and switches for power accessories. Black cut-pile carpeting was used. *(McAtee)*

The individual rear bucket seats had the same basket-weave pattern as the fronts. They were divided by a black-carpeted separater which was decorated with a 300 medallion. A combination armrest and storage console was provided. *(McAtee)*

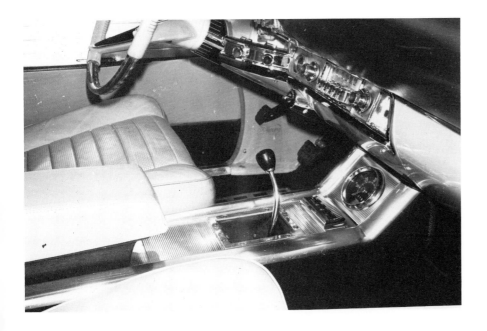

Some of the four-speed F's were delivered with a curved type of shift lever, like this one in the extremely rare Cone/Hoover convertible. The shifter protruded through the console in place of the ashtray. A console-mounted tachometer was standard.

TorqueFlite-equipped cars had a sliding-top ashtray built into the center console up front. A tachometer was also provided with the automatic gearbox and some Chrysler engineers felt it was extraneous. *(Hoover)*

The dashboard used on the 300-F was straight from the Buck Rogers's school of automotive design. Engine monitoring gauges were housed in a semi globular display bubble, which also incorporated the 150-mph speedometer. *(McAtee)*

Bob McAtee's car wears a chrome engine dress-up kit. According to the marque experts who previewed this book, this kit was not offered until 1964, but could be dealer retro-fitted to earlier cars. McAtee, however, showed the author a factory service bulletin which seemed to indicate this feature was a *Chrysler* dealer-installed accessory at an earlier date. This dress-up kit is not listed specifically as a Letter Car accessory, however.

On stick-shift cars a horizontally ribbed plate was used to cover the housing for the TorqueFlite controls. The toggle switch in the center of the plate operated the backup lamps. The lever below the toggle controlled the turn signals. *(Cone/Hoover)*

CHAPTER EIGHT
1961 Chrysler 300-G
Grand Touring Car...
Letter Style

In November 1959, Chrysler Division undertook an intensive market survey aimed at pinpointing the typical buyer of a Chrysler 300 Letter Car. It was discovered that the majority of owners were successful executives and businessmen who liked to drive fast. An interesting statistic revealed by the report was that the median distance these drivers traveled each year was over 15,000 miles. At the time, this was nearly twice the national average of all American car owners. In addition, most of those participating in the survey indicated that appearance was the primary motivation for purchasing a 300. Performance and handling were picked as the second most important factor.

This had a strong effect on the company's planning for the 1961 model year. A significantly revised promotional program was designed to place stress on the Letter Car as a *grand touring* automobile. The sales folder, for example, depicted the new 300-G hardtop and convertible against settings ranging from Sutton Place, in New York City, to Acapulco, Mexico. According to the copy, the newest contender in a family tree of rugged race-bred automobiles was "Styled in

The last Milestone Letter Car was promoted as a Grand Touring automobile and the sales folder stressed this idea with sketches of hardtops and convertibles in different locales. They seemed to indicate a trip from New York to Aspen, Colorado, and then down to Mexico. Ed and Bev Aldridge are likely to be seen on tour, too, behind the wheel of this Mardi Gras Red hardtop.

the tradition of its clean-lined lineage. And by every high-performance standard, ready to prove its worth as America's most luxurious 'Grand Touring' Automobile." Quotes from Tom McCahill and Brewster Shaw alluded to the performance and handling characteristics of the G.

This concept of the Chrysler 300 Letter Car as a long-distance traveling machine would soon start to eclipse the traditional image of a race car built for the street. By 1962 a non-letter-edition Chrysler 300 (essentially a Windsor with sportier features) would appear on the market and dramatically dilute the prestige of the model nameplate. True Letter Cars would remain in production through 1965, but the 300-G marked a turning point in the history of the line. It is, today, the last of the letter series models recognized as a Milestone Car. Some of the editions that followed would be equal or superior to the G in terms of performance, handling, top speed, style, luxury or rarity; but none would enjoy the distinction of being a *unique* offering within the Chrysler model range. The 300-G was also the last of the series to wear high tailfins and the personalized touch of designer Virgil Exner.

Appearance changes for the 300-G were of the minor variety. The quad headlights were switched to a diagonally vertical mounting and the grille was changed to an inverted trapezoid shape. Many enthusiasts came to think of this as the "smiling" grille, as compared to the "frowning" style used the year before. There were also new taillights, a redesigned rear bumper, a ribbed deck lid without a spare tire impression and some small trim and ornamentation revisions. Interiors were virtually the same as in 1960, except for a difference in the basket-weave pattern embossed on the four bucket seats.

There were two major technical changes on the last Milestone Chrysler 300. The expensive Pont-à-Mousson transmission was no longer available, although the 400-hp engine remained. It was now provided with a heavy-duty three-speed manual gearbox controlled via a floor-mounted shifter. This was an awkward unit having a non-Synchromesh first gear and could not match the acceleration provided in the

Resplendently trimmed in Formal Black attire with a sporty white nylon "top hat" is Evaline McAtee's 300-G convertible, an original car. The 300-G featured styling refinements to the headlights, taillights and rear deck lid, plus a few minor technical changes.

Size 8.00x15 Goodyear Blue Streak high-performance nylon super-cushion tubeless tires of six-ply construction were standard, as were ventilated wheel covers with whitewalls. A circular hub center with three short spokes housed a medallion. *(McAtee)*

TorqueFlite automatic. Consequently, very few cars were sold with this attachment. The second mechanical revision was the use of an alternator with *both* 1961 engines. (This feature had been used exclusively with the 400-hp V-8 in the previous season). The new electrical system lessened the battery drain on highly accessorized cars, since charging took place even when the car was idling. In addition, service life of the alternator was said to be three times longer than with a direct-current generator.

Handling was also improved a bit on the G by a switch to larger tires and fifteen-inch rims. The primary advantage of the bigger wheels was better brake cooling, but the new 8.50x15 size tires provided better traction during driving and acceleration, too.

Sales of the 300 series improved significantly this season, despite management problems within Chrysler Corporation. Tex Colbert had resigned in April 1960, passing the office of president to William C. Newberg, his hand-picked successor. Three months later, Newberg also stepped down, due to a scandal involving conflicts of interest from connections with supplier firms. Colbert resumed control, but an embittered Newberg launched corporate warfare and somewhat disrupted day-to-day operations of the company. At a stockholder's meeting in April 1961, Colbert was directly challenged by Newberg's supporters. This situation continued and, three months later, Colbert resigned for his second and last time. He was replaced by Lynn A. Townsend.

Virgil Exner followed Colbert in departing the company and soon gave up his posts as vice president and chief of styling. He would continue with the corporation until 1964, but his days of direct involvement in designing new cars were over. Thus, the 300-G warrants some additional recognition as the last model in the letter series to bear full benefit of the master's touch. The 1962 300-H, although reflecting some Exner influences, would not conform to his original plans for that model year.

It goes without saying that both of these men had played major roles in Chrysler 300 history. Colbert by supporting the hemi's development and the engineering program that gave us push-button gear-shifting, torsion-bar suspension and ram-tuned induction; and Exner by designing the original C-300 and all of the

Outside rearview mirrors remained an option and were of a new design this year. Electric wipers and automatic washers were standard. Tinted glass, including a shaded rear window, was optional. *(Aldridge)*

The tall tailfin look was retained as illustrated on this V-8-powered G. The contour of the finned section had a horizontally elongated V-shape. The fins and fenders were angled on intersecting planes and formed a crease contour down the center of the "vee." Rear overhang was 59 inches. (Tires shown in photo are not original.) *(Krause Publications)*

classic Letter Cars that followed it. Thus, it seems justly fitting that this history of the Milestone Chrysler 300's ends in the same year that both gave up active involvement with the series.

As for the racing and performance record of 1961, there was minimal activity. Gregg Ziegler returned to Daytona driving a new 300, which Bob McAtee helped him purchase. He took Flying Mile honors with a run of 143.00 mph. A second 300 pilot, Harry Faubel, Jr., set the pace in the Standing Start Mile competition with an average speed of 90.7 mph. According to contemporary road testers the G could go from 0-60 in 8.4 seconds and from 0-100 in twenty-one. Quarter-mile times were in the sixteen-second range with a terminal speed of just under 90 mph. *Consumer Reports* bluntly stated that the 300-G was "too big," but still referred to it as "an unskimped and solid hell-for-leather car." In short, the 1961 model marked a turning point in the history of the letter series, but it was still a pure-blooded *enthusiast's* machine.

SPECIFICATIONS
Model number   300-G (Chrysler RC4-F)
Serial number data   No longer applicable. After 1959 the serial number
Engine number data   sequence for *all* Chrysler models was the same, with first four digits added, as applicable, to identify the model. All "300" numbers began with 8413, but the six-digit number indicating production sequence was not exclusive to the 300 series.
Body style numbers   (hardtop) 842 (convertible) 845
Announcement date   January 1961
Advertised dealer price   (hardtop) $5,413.00 (convertible) $5,843.00
Standard equipment
Special wedge engine, heavy-duty suspension, custom steering wheel, Safety-Cushion dash panel, tilt-type rearview mirror, nylon high-performance white sidewall tires, electric clock, directional signals, power brakes, power steering, hand-brake warning signal, TorqueFlite transmission, chrome stainless steel wheel covers, windshield washer, undercoating, hood insulation pad, leather upholstery, canted dual headlamps, Silent-Flite fan drive (limits fan speed to 2500 rpm), swivel seats, low back-pressure exhaust system, power windows, tachometer, center armrests (front and rear).

The contour of the front wheel opening was extended forward to the top inner corner of the parking lamp, with a parallel crease running forward to meet the upper outer corner of the lens. The result was a rectangular indentation. *(Aldridge)*

The concave panel at the rear body was trimmed by a chrome outline molding. The bumper was a fairly straight design with two small rubber-faced bumperettes in the center. They were spaced to surround the license plate. *(Aldridge)*

Options and accessories
(Code)—description—price
- (311)—air conditioning, unit w/heater—$510.40
- (313)—air conditioning, unit w/accessory group #306—$408.50
- (324)—power-operated radio antenna—$25.90
- (617)—permanent antifreeze—$5.90
- (387)—crankcase vent system (required on California cars)—$5.20
- (339)—rear window defogger (hardtop only)—$20.85
- (298)—vacuum door locks—$36.60
- (379)—Sure-Grip differential—$51.70
- (315)—Custom Conditionaire heater—$101.90
- (383)—left-hand outside mirror, remote control—$18.00
- (293)—power front seats—$101.90
- (422)—Golden-Tone radio—$99.80
- (427)—Golden-Tone radio with rear speaker—$117.10
- (423)—Golden Touch-Tune radio—$124.10
- (428)—Golden Touch-Tune radio with rear speaker—$141.40
- (429)—rear shelf radio speaker (hardtop only)—$17.30
- (441)—Solex tinted glass (all windows)—$43.10
- (***)—three-speed manual transmission with 375-hp V-8—price not available
- (***)—400-hp V-8 (three-speed manual transmission required)—price not available

Accessory groups:
- (306)—Golden Touch-Tune radio with rear speaker, heater, rear window defogger, power antenna and accessory package A—$302.55
- (301)—Accessory package A—door edge protectors, rear license plate frame and vanity mirror—$10.50

Original paint colors
(factory code)	name	Ditzler combination
(BB-1)	Formal Black	9000
(PP-1)	Mardi Gras Red	71203
(RR-1)	Cinnamon	71140
(WW-1)	Alaskan White	8218

Interior trim
Standard upholstery color   Tan (beige)
Standard upholstery material   Smooth-grained natural tan (beige) cowhide, basket-weave textured leather, black cut-pile carpets
Seat welting, color/type   Tan vinyl
Dashboard color   Black Frost metallic gloss
Door panels   Tan vinyl w/brushed-aluminum inserts

Convertible top
Color   Black or white
Material   Nylon fabric
Notes:
Code 999 on vehicle data plate indicates special-order paint combination
Code 888 on vehicle data plate indicates special-order interior combination
Engine
Same as 1960 without change
Valve specifications
Same as 1960 without change
Carburetion
Same as 1960 without change
Electrical
Original battery type   Willard H-12-70 or Auto-Lite 12-H-79, 12-volt negative ground
Original alternator   Chrysler 1889200
Standard output   35 amperes
Heavy-duty output   40 amperes
Transmission
(TORQUEFLITE)
Type   three-speed automatic, torque converter with planetary gears
Ratios   (low) 2.45:1 (drive) 1.45:1 and 1.0:1
Control location   push-button, left side of dash

The rear flanks were again decorated with spear moldings and a large medallion. The spears were somewhat shorter and did not sweep across the rear panel. Chrysler Unibody construction included 140 pounds of sound-deadening spray and 100 square feet of padding to cut vibration and noise. *(Aldridge)*

A short chrome riser molding decorated the trailing edge of the central deck lid rib and formed a trunk handle. It crossed over a circular model medallion. The Chrysler name, in block letters, was at the lower right corner. *(Aldridge)*

(MANUAL)
Type   Chrysler three-speed manual
Ratios   (first) 2.55:1 (second) 1.49:1 (third) 1.00:1
Control location   floor-mounted lever
Clutch diameter   11 inches

Running gear
Differential   hypoid type, semi-floating
Axle gear ratios   (standard) 3.23:1 (optional) 3.58:1, 3.15:1 and 3.73:1
Standard power steering   Chrysler Constant-Control type, gearshaft-and-sector design with power piston
Standard power brakes   Chrysler Total-Contact type

Chassis and body
Frame type   Bolted-on front stub frame
Body construction   welded steel, unitized body
Wheelbase   126 inches
Front tread   61.2 inches
Rear tread   60.0 inches
Overall length   219.8 inches
Overall height (loaded)   (hardtop) 55.6 inches (convertible) 56 inches
Overall width   79.4 inches
Ground clearance   6.3 inches
Shipping weight   (hardtop) 4,260 pounds (convertible) 4,315 pounds

Suspension
Same as 1960 without change

Wheels and tires
Tire size   8.00x15
Whitewall width   2-7/8 to 3 inches
Standard disc wheels   15x6K

Production totals
Two-door hardtop   1,280
Two-door convertible   337
Total   1,617
Remaining 1981*   251 (15.5 percent)
*See Chapter Two

Sales literature
1—Catalog No. CS-446, undated, 12 pages, black & white, 3½ x 10 inches. Titled: *Chrysler 300 G.*
2—Sheet No. (C), September 1960, 3 pages, printed by Ross Roy, Inc. Specifications, accessory list, price comparisons, all cars.

Advertising
1—*Motor Trend,* December 1960
2—*Road & Track,* January 1961, back cover
3—*Sports Car Illustrated,* January 1961, 300-G ad

The door panels again had a beige (tan) vinyl covering, brushed-aluminum insert panels, Chrysler 300 model designations and a band of black carpet along the bottom edge. *(Aldridge)*

Regular interiors were of tanned, natural grain cowhide with black carpet inserts on the rear of the front seat backs. Seat center inserts used a perforated basket-weave embossment with small squares arranged in rows of four. *(Aldridge)*

Rear bucket seats were divided by the console and an upholstered armrest that lifted to reveal a storage compartment. The area between the backrests was black-carpeted and decorated with a model medallion. *(McAtee)*

The instrument panel layout was nearly identical to that of the 300-F. A black pad with wide-spaced pleats covered the top of the dash. A non-glare inside rearview mirror was standard. *(Aldridge)*

The center console housed a tachometer, sliding ashtray and power window controls in the front section. The rear portion of the console incorporated a cigarette lighter, ash receiver and two switches for the rear window lifts. *(McAtee)*

The entire dash panel of the 300-G was finished in Black Frost Metallic gloss. Sahara Tan finish was no longer used. An electric clock was standard equipment and the large locking glovebox lid was decorated with a Chrysler script. *(Aldridge)*

CLUBS

The following club listing is of organizations that may be of interest to the 300 Letter Car enthusiast. The clubs should be contacted individually for up-to-date information about their publications, member service, requirements and dues.

AIRFLOW CLUB OF AMERICA
4927 Wyoming Avenue
Harrisburg, Pennsylvania 17109

ANTIQUE AUTOMOBILE CLUB
OF AMERICA
501 West Governor Road
Hershey, Pennsylvania 17033
717-534-1910

CHRYSLER 300 CLUB, INC.
111 Melrose
Texarkana, Texas 75503

CHRYSLER 300 CLUB INTERNATIONAL, INC.
19 Donegal Court
Ann Arbor, Michigan 48104
313-971-3274

CLASSIC CAR CLUB OF AMERICA, INC.
Post Office Box 443
Madison, New Jersey 07940
201-377-1925

DAYTONA AND SUPERBIRD ASSOCIATION
Post Office Box 1100
Cerritos, California 90701
213-860-8320 or
714-523-1109

IMPERIAL OWNERS CLUB
Post Office Box 991-CHS
Scranton, Pennsylvania 18503

THE MILESTONE CAR SOCIETY, INC.
Post Office Box 50850
Indianapolis, Indiana 46250

MOPAR MUSCLE CLUB
4910 Leigh
Amarillo, Texas 79110

NATIONAL HEMI OWNERS ASSOCIATION
7010 Darby Avenue
Reseda, California 91335
213-344-5531 or
213-345-0314

W.P.C. CLUB, INC.
P.O. Box 4705
North Hollywood, California 91607

RESTORATION AIDS

The Chrysler 300 lover will find two basic types of restoration aids. First, sources of parts and literature. Second, firms that provide services, such as rebuilding a carburetor or restitching a seat. Therefore, this list of restoration aids is broken-up into categories that cover a variety of old car hobby parts or service providers. However, it must be remembered that some of these companies or individuals deal with the restoration of numerous types of cars.

Axles
Sommers Bros.
530 S. Mountain Ave.
Ontario, CA 91761

Batteries
Keystone Corp.
35 Holton St.
Winchester, MA 01890

Brake Components
Robert Midland
Box B
Hawley, PA 18428

Carburetors
Douglas Mc Roberts*
7380 Cannel Dr
Boulder, CO 80301

Chrysler 300 Club, Inc.*
111 Melrose
Texarkana, TX 75503

Chrysler 300 Club Int'l.*
19 Donegal Ct.
Ann Arbor, MI 48104
(rebuild kits)

James Alexandro
P.O. Box 144
Maspeth, NY 11378

The Carb Shop
Route 1 Box 230-A
Eldon, MO 65026

Chrome Plating
Graves Plating Co*
P.O. Box 1052
Industrial Park
Florence, AL 35633

Convertible Tops
Paul Weissman
48 Appleton Rd
Auburn, MA 01501

Decals
Bob Austin
6925 S.W. 73rd Ct.
Miami, FL 33143

Gary Goers*
6738 Lemora
Pico Rivera, CA 90660

Engine Parts
Ed Hamburgers*
Hi-Perf Parts
1590 Church Rd.
Toms River, NJ 08753

Dennis Cloer*
Rt 3 Box 213
Newton, NC 28658

Merle Wolfer*
6641 Parkway
Gladstone, OR 97027
(Ram Tube Connectors)

Chrysler 300 Club, Inc.*
111 Melrose
Texarcana, TX 75503

Chrysler 300 Club Int'l.*
19 Donegal Ct.
Ann Arbor, MI 48104

Engine Parts
George Riehl*
Mopar Dyno-Saurs
19 Donegal Ct.
Ann Arbor, MI 48104
(Engine Rebuilding)

Lunati Cams*
3871 Watman Ave.
P.O. Box 18201
Memphis, TN 18201
(Crankshaft Repairs)

Ed Iskandarian Racing Cams*
16020 South Broadway
Gardena, CA 90248

Egge Machine Co.
8403 Allport
Santa Fe Springs, CA 90670

Exhaust Parts
Dave Schwandt
P.O. Box 233
Earlville, IA 52041

Jim Fortin
95 Weston St.
Brockton, MA 02461

Roberts Brass*
P.O. Box 27
Cowanesque, PA 16918
(1960-61 exhausts)

Windshields/Glass
Burchingers K-F Parts
P.O. Box 661
Chicago, IL 60666

Classic Glass
2100 Prater Way
Reno, NV 89431

Hardware & Fasteners
Guy C. Close
13426 Valna Dr.
Whittier, CA 90602

H.C. Fasteners
P.O. Box 817
Alvarado, TX 76009

Restoration Specialties
P.O. Box 445
Somerset, PA 15501

Ignition Parts
Curtis Ackerman, Jr.
West Burlington, NY 13482

Bob Futterman
722-B Carswell Ave.
Holly Hill, FL 32017

Standard Motor Products
37-18 Northern Blvd.
Long Island City, NY 11101

Instrument Repairs
Lu Berger*
Route 2 Box 53-F
Mountain Home, AR 72653
(clocks)

Bobs Speedos & Clocks
15255 Grand River
Detroit, MI 48227

Reynolds Repairs
4 Loboa Dr.
Danvers, MA 01923

Lighting
American Arrow
625 Redwood Dr.
Troy, MI 48084

Headlight Headquarters
35 Tinson St.
Lynn, MA 01902

Arnold Levin
2835 W. North Shore Ave.
Chicago, IL 60645

Literature
Wisconsin Chapter*
Chrysler 300 Club, Inc.
Ken Block
Box 7052
Rochester, MN 55903
(carb spec sheet reprint)

Chrysler 300 Club, Inc.*
111 Melrose
Texarcana, TX 75003
(owners manual & shop manual reprints)

Steve Soloy*
2266½ Torrance Blvd.
Torrance, CA 90503

Classic Motorbooks
Rt 1
Osceola, WI 54020
(1950-65 Chrysler Parts Interchange Manual)

Chrysler 300 Club, Int'l.*
19 Donegal Ct.
Ann Arbor, MI 48104

Locks
Bills Lock Shop
3605 Robbin Rd
Toledo, OH 43623

Non-Automotive 300 Accessories
Chrysler 300 Club*
International, Inc.
19 Donegal Ct.
Ann Arbor, MI 48104 (See Clubs)

Chrysler 300 Club, Inc.*
111 Melrose
Texarcana, TX 75003

McHenry Trophies*
3715 W. John St.
McHenry, IL 60050
('300' Den Clock)

Paint Specialist
Bob Drew*
10425 Pinyon Ave.
Tujunga, CA 91042

Radios
Antique Radio Doctor
Poplar Hill Rd.
W. Lebanon, ME 04027

Restoration Parts
George Cone*
701 N. Lillian St
McHenry, IL 60050

Denis Cloer*
Rt 3 Box 213
Newton, NC 28658

Rubber Parts
Gary Goers*
6738 Lemora
Pico Rivera, CA 90660

Floating Power*
Motor Mount Rebuilding
RR 1 Box 332
Huntington, IN 46750

The Brentwood Co.
P.O. Box 761
Brentwood, TN 37027

Metro Molded Parts
3031 2nd St.
Minneapolis, MN 55411

Lynn Steele
Rt 1 Box 71
Denver, NC 28037

Wefco Rubber Mfg.
1655 Euclid Ave.
Santa Monica, CA 90404

Reproduction Parts
Gary Goers*
6738 Lemora
Pico Rivera, CA 90660
(300 interior parts and complete interior kits)

Charles Fabian*
4020 S. Michigan
Chicago, IL 60653
(300 wire wheel caps)

Jerry Schlegel*
1910 "A" St.
Forest Grove, OR 97116
(300 H grille badge)

John Sheets*
101 Ridge Rd.
Horseheads, NY 14845
(300 H grille badge)

Ray Beaumont*
Rt 3 Box 350C
Callahan, FL 32011
(300 F-G grille badge)
(1957-61 weather strip)

Bob Austin*
6925 S.W. 73rd Ct.
Miami, FL 33143
(Decals 300 C-D-E-H)

George Riehl*
19 Donegal Ct.
Ann Arbor, MI 48104
(1957-59 grille badge)
(1957-59 glove box badge)
(1957-59 trunk medallion)

G&H Metal Finishers*
282 Dakota St.
Paterson, NJ 07503
(300-F-G-H floor trim refinishing)

Auto Body Reproductions
8792 Quigley St.
Westminster, CO 80030

Taillight Lenses
Chrysler 300 Club, Inc.*
111 Melrose
Texarcana, TX 75003

Chrysler 300 Club Int'l.*
19 Donegal Ct.
Ann Arbor, MI 48104

Transmission
The Klocko*
1023 Edgewater Lane
Ingleside, IL 60041
(Repair Specialist)

George Riehl*
19 Donegal Ct.
Ann Arbor, MI 48104
(parts & rebuilding)

Denis Cloer*
Rt 3 Box 213
Newton, NC 28658
(Parts)

Fatsco Transmission Service*
Rt 46
Fairfield, NJ 07006

Mike Meier
511 First St. N.
Virginia, MN 55792
(Rebuilding)

Errol McIntire
202 N. Fourth St.
Woodsfield, OH 43793
(rebuilding)

Tires (Specialty)
Kelsey Tire Co.
P.O. Box 564
Camdenton, MO 65020

Lucas Automotive
2850 Temple Ave.
Long Beach, CA 90806

Universal Tire Co.
2650 Columbia Ave.
Lancaster, PA 17603

Upholstery
Gary Goers*
6738 Lemora
Pico Rivera, CA 90660

Ray Simcuski*
119 Fisco Ave.
Syracuse, NY 13205

*Indicates have specialized knowledge or skills that apply mainly to Chrysler 300 restoration work.

INFORMATION AND SOURCES

The Chrysler 300 Letter Car enthusiast will find a lot of company in the old car hobby. There are many articles and books to help him learn about the history and technical details of the car he owns. In addition numerous stories have been printed in the *300 Club News* and *Brute Force*, two excellent publications put out by the two marque clubs.

Membership in one or both of these hobby organizations is highly recommended for every Chrysler 300 lover.

The following represents an extensive bibliography of commercially published references to the Milestone Letter Cars. Contributors to the bibliography include Ray Doern of the Chrysler 300 Club, Inc., and James Banach of the Chrysler 300 Club International, Inc. The major portion of the list appeared in *News-Flite*, a newsletter published by the latter organization and is used with the permission of the *News-Flite* editors.

Antique Motor News
9/73 — 300 article — pg. 12

Auto Age
6/55 —
7/55 — C-300 test — pg. 11
4/56 — picture — pg. 11

Automobile Quarterly
Vol. 3 No. 8 — 1976 — 300 history

Autoweek
6/9/73 — 300 history — pg. 19
6/16/73 — 300 history — pg. 20

Basic Ignition & Electrical System

Hot Rod
No. 2 — 1971 — 300-F picture — pg. 164

Car Classics
10/74 — 300 history — pg. 36

Car Craft
12/60 — 300-F picture — pg. 17

Car and Driver
8/61 — 300 history — pg. 19
9/61 — 300 history — pg. 60
10/61 — 300 history — pg. 30
1/62 — 300-H ad — pg. 81
6/74 — C-300 feature — pg. 54

Car Exchange
8/80 — 300 history — pg. 50
9/81 — 300-F — pg. 50

Car Life Annual
1955 — C-300 feature — pg. 46
1955 — C-300 engine — front cover

Cars
5/60 — 300-F test — pg. 23
5/60 — 300-G test — pg. 16
5/70 — 300-D test — pg. 46
5/70 — 300 history — pg. 29

Car Life
12/59 — 300-E test — pg. 67
3/60 — 300-F report — pg. 22
8/61 — 300-G test — front cover, pg. 14
6/62 — C-300 picture — pg. 53
8/70 — 300-C feature — pg. 40

Cars & Parts
9/72 — 300 history — pg. 88
10/11/72 — 300 history — pg. 84

Cavalier
6/75 — 300-F picture — pgs. 62-63

Chilton's Motor Age
10/55 — C-300 picture — pg. 21

Chrysler Performance Handbook
1962 — 300-F feature — pg. 90

Classic Car Quarterly
Fall 1979 — 300 history — pg. 32

Golden Lions Newsletter
8/71, 10/71, 12/71 — 300 history
2/72, 4/72, 6/72, 8/72 — 300 history

Hot Rod
2/56 — C-300 picture — pg. 41
5/56 — 300-B picture — pg. 18
9/56 — 300-B test — pg. 26
11/56 — 300-B picture — pg. 49
5/57 — 300-C picture — pg. 29
11/57 — C-300 picture — pg. 30
2/59 — 300-D picture — pg. 99
3/59 — 300-D picture — pg. 99
4/60 — 300-F test — pg. 36, F.C.
12/60 — 300-F picture — pg. 17
5/61 — 300-F, G picture — pgs. 46, 113
12/61 — 300-F picture — pg. 34

How-To-Hotrod
1961 — 300-F — pg. 132

Mechanix Illustrated
5/55 — C-300 test — pg. 98
12/55 — C-300 picture — pg. 105
6/56 — 300-B test — pg. 90
12/56 — 300-B picture — pg. 97
5/57 — 300-C test — pgs. 82, 94
11/57 — 300-C picture — pg. 82
5/60 — 300-F test — pg. 90

Milestone Car
No. 15 — Spring 1976 — 300 feature

Motor
2/55 — C-300 report — pg. 139
2/56 — 300-B picture — pg. 59

Motor Life
4/55 — C-300 report — pg. 10
5/55 — C-300 test — pgs. 12, 22
6/55 — C-300 report — pg. 20
12/55 — C-300 racing — pg. 11
5/56 — 300-B report — front cover, pgs. 27, 49
9/56 — 300-B report — pg. 14
12/56 — Bendix EFI — pg. 7
1/57 — 300-C report — pg. 64
5/57 — 300-C report — pgs. 40, 46
5/58 — 300-D test — pg. 62
3/59 — 300-E picture — pg. 81
2/60 — 300-F report — pg. 68
4/60 — 300-F report — pg. 48
1/60 — 300-F test — pg. 48
7/60 — 300-F report — pg. 42

Motorsports
5-6/55 — C-300 picture — front cover
7-8/55 — C-300 test — pg. 12
4/61 — 300-G picture/test — front cover, pg. 24

Motor Trend
5/55 — C-300 picture — pg. 30
6/55 — C-300 report — pg. 52
9/55 — C-300 picture — pg. 42
11/55 — C-300 picture — pg. 29
8/56 — 300-B picture — pg. 29
12/56 — Bendix EFI — pg. 56
3/57 — 300-C report — front cover, pg. 40

4/57 — 300-C picture — pg. 9
5/57 — 300-C report, EFI — pg. 59
11/57 — fuel injection — pg. 29
12/57 — Bendix EFI — pg. 21
2/58 — 300-D test — pg. 28
6/58 — 300-D picture — pg. 58
2/59 — 300-E feature — pg. 12
3/59 — 300-E report — pg. 58
11/59 — ram-induction — pg. 45
2/60 — 300-F test — pg. 22
6/61 — 300-G test — pg. 52
12/61 — 300-B, F, G pictures — pg. 42

Old Cars Weekly
11/18/75 — history — pg. 10

Plymouth/Dodge/Chrysler Handbook
1973 — 300 history — pgs. 17, 72, 172

Playboy
4/74 — C-300 feature — pg. 156

Old Cars Illustrated
3/79 — 300-B photos — pg. 40, back cover

Popular Mechanics
2/56 — C-300 racing, 300-B pic. — pg. 188, front cover
12/56 — C-300 picture — pg. 80
2/57 — 300-C feature
2/58 — 300-D picture — pg. 161
1/59 — 300-E picture — pg. 164

Popular Science
4/55 — C-300 test — pg. 131

Road & Track
6/55 — C-300 test — pg. 28
4/58 — 300-D test — front cover, pg. 45
3/60 — 300-F feature — pg. 64

Rod Builder & Custom
6/57 — 300-C picture — pg. 10

SAE Transactions
Vol. 65-1957 — 300-D EFI — pg. 758

Science & Mechanics
8/56 — 300-B test — pg. 95

Special-Interest Autos
6/79 — C-300 test — front cover, pg. 8

Speed Age
6/55 — C-300 picture — front cover
6/55 — C-300 racing — pgs. 8, 24
8/55 — C-300 feature — front cover, pg. 36
9/55 — C-300 racing — pg. 19
12/55 — C-300 racing — pg. 14
1/56 — C-300+test — pgs. 13, 16, 20, 23
3/56 — C-300 feature — pgs. 29, 52, 49
4/56 — C-300 racing — pg. 22
7/56 — 300-B report — front cover, pg. 18
9/56 — 300-B racing — pg. 60
3/57 — 300-C report — front cover, pg. 36
3/57 — 300-B report — pg. 59
3/59 — 300-E test — front cover, pg. 32

Spectator (Chrysler Corp.)
Winter 1979 — pictures & history

Speed Mechanics
6-7/55 — C-300 picture — pg. 19

Sportscar Graphic
6/60 — 300-F test — front cover, pg. 64

Sports Car Illustrated Directory
1958 — 300-C test — pg. 78

Sports Car Illustrated
4/56 — 300-B test — pg. 8
8/59 — 300-E test — pg. 26
1/61 — 300-G

True's Automobile Yearbook
1957 — racing, Carl Kiekhaefer

WPC Club News
10/77 — 300-B — pg. 4
12/80 — 300-H — pg. 4

BOOKS

Chrysler & Imperial: The Postwar Years, by Richard Langworth
Seventy Years of Chrysler, by George Dammann
Encyclopedia of American Cars 1940-1970, by editors of Consumer Guide
Encyclopedia of American Cars 1946-1959, by James Maloney & George Dammann

INDEX

Agabashian, Fred, 7
Arnold, Bill, 7

Baker, Buck, 16, 24
Baldwin, Maury, 28
Beauchamp, Bert, 56
Bendix Corporation, 39-40
Bettenhausen, Tony, 9, 11
Breer, Carl, 7
Brute Force, 40

Callahan, Bill, 48
Car Life, 34, 61
Carlson, Billy, 7
Carroll, Bill, 30
Carter carburetors, 15, 18-19, 25, 33, 43, 50, 60
Colbert, L. L., 37
Colbert, Tex, 54, 67
Consumer Reports, 48, 68
Cunningham, Briggs, 8-10, 30
Cunningham, Gil A., 40
Curtiss-Wright, 54

DePalma, Ralph, 6-7
Dice, Dick, 41

Eames, Danny, 57-58
Ewert, Harry, 48
Exner, Virgil, 13, 28, 34, 38, 66-67

Faubel, Harry, Jr., 57-58, 68
Firestone Tire and Rubber Company, 11
Fitch, John, 9-10
Flock, Tim, 16-17, 23, 58

Goodyear, 19, 26, 34, 44, 47, 51, 61, 66
Granatelli, Andy, 58

Harroun, Ray, 6
Hartz, Harry, 7
Hilborn fuel-Injection, 11, 55
Holley, 33

Howie, George N., 7
Howley, Tom, 16

Kahn, Bernard, 58
Kalenberg, Sherwood, 40
Kelsey-Hayes, 11-12, 14, 18-19
Kiekhaefer, Carl, 9-10, 16-17, 19, 23-24, 57
Kirkwood, Pat, 10
Koechling, Warren, 17, 57-58
Kurtis chassis, 11

Life, 51

Mallory ignition systems, 9
McCahill, Tom, 9, 66
McFee, Roger, 10
Mechanix Illustrated, 9, 19, 34
Mercury Outboard Company, 9, 16, 23-24
Meyer-Drake, 11
Motor Life, 19, 34
Motor Trend, 14, 19, 34, 40 47-48, 61, 70
Mundy, Frank, 17

National Geographic, 51
New Yorker, The, 44
Newberg, William C., 67
News-Journal, 58

Pemberton, Bob, 49
Petty, Lee, 6, 8, 11
Piasecki Aircraft Company, 54
Poirier, Tom, 13
Pont-a-Mousson transmission, 54-57, 66
Popular Mechanics, 19
Popular Science, 34

Quinn, Ed, 12-13

Register, 19
Rickenbacker, Eddie, 6
Road & Track, 34, 70
Rodger, Bob, 11-13, 19
Rogers, Buck, 63
Ross Roy, Inc, 19, 31
Rouse, Larry, 57, 58

Shaw, Brewster, 11, 41, 57, 58, 66
Skelton, Owen, 7
Solex, 13, 18, 32, 42, 50, 59, 69
Special-Interest Autos, 16
Spectator, 13
Speed Age, 49
Sports Car Illustrated, 19, 48, 61, 70

Thatcher, Norm, 41
Thompson, Speedy, 16
Time, 51
Townsend, Lynn A., 67

Voss, Cliff, 13

Walters, Phil, 10
Ward's Automotive Yearbook, 54
Willard, 18, 43, 51, 53
Wood, Vicki, 39, 49

Zeder, Fred M., 7
Zenith carburetors, 9
Ziegler, Gregg, 41, 48, 49, 56-58, 68